Bibliografische Information der Deutschen Nationalbibliothek

Die Deutsche Nationalbibliothek verzeichnet diese Publikation in der
Deutschen Nationalbibliografie; detaillierte bibliografische Daten sind
im Internet über http://dnb.d-nb.de abrufbar.

ISBN 978-3-8325-3501-8

Logos Verlag Berlin GmbH
Comeniushof, Gubener Str. 47,
10243 Berlin
Tel.: +49 (0)30 42 85 10 90
Fax: +49 (0)30 42 85 10 92
INTERNET: http://www.logos-verlag.de

Modular Ontologies for Spatial Information

Joana Hois

A thesis submitted for the degree of
Doktor der Ingenieurwissenschaften (Dr.-Ing.)

Die vorliegende Dissertation wurde dem Fachbereich 3 (Mathematik und Informatik) der Universität Bremen im November 2012 vorgelegt. Die Arbeit entstand im Rahmen des Sonderforschungsbereichs SFB/TR8 Spatial Cognition, Teilprojekt I1-[OntoSpace].

Mitglieder des Prüfungsausschuss waren Prof. John A. Bateman, PhD (Erstgutachter), Prof. Dr. Kerstin Schill (Zweitgutachter), Dr. John D. Kelleher, Prof. Michael Beetz, PhD, Dr. Oliver Kutz, Gerrit Bruns. Das Promotionskolloquium fand am 22. Mai 2013 statt.

Abstract

This thesis investigates the development and application of spatial ontologies. It also analyzes methodologies and techniques necessary to model spatial ontologies and to apply them in different application scenarios.

First, we therefore analyze the variety and diversity of spatial information. We analyze in detail, which types and conceptualizations are available to model spatial information. In particular, we distinguish different types of spatial information by specifying their thematically different perspectives on space. Based on these perspectives, spatial information that is characterized by quantitative, qualitative, abstract, domain-specific, or multimodal aspects is distinguished. Different perspectives guide the development of computational models for spatial information as ontology modules. Ontologies are logic-based specifications that categorize and classify semantic types and relations. To model the spatial ontologies, we technically use OWL and Protégé as well as CASL and HeTS.

Second, we investigate how the developed ontologies can be compared and combined, technically as well as contentwise. Here, ontological modularity is essential to provide combination methods for the different spatial ontologies. We discuss and apply primarily the methods extension/refinement, matching, and connection to combine ontology modules that comply with different spatial perspectives. The use of these combination methods between ontologies is determined by means of their technical practicability and thematic adequacy. The combination of spatial ontology modules is applied in different application scenarios.

Third, we integrate uncertain information with regard to the developed ontology modules. As spatial applications have to cope with partially vague, uncertain, or ambiguous types of information, we provide a method to take these aspects into account in the ontological modeling of spatial information. No standards are available today that support uncertain types of information together with logically strict ontological formalizations. Hence, we examine technical and thematic approaches to combine uncertainties and ontologies and to use them in particular application scenarios.

Finally, the developed spatial ontology modules, their combinations, and their uncertainty components are applied and evaluated in three applications: in the area of architectural design and assisted living systems, spatial ontology modules are applied to provide automated assistance and to analyze design requirements; in the area of visual recognition, spatial ontology modules are used to support object and scene classifications; in the area of natural language processing, spatial ontology modules are used to interpret natural language in terms of qualitative spatial representations.

In conclusion, spatial ontology modules are developed in this thesis that are structured by means of their thematically distinct perspectives on space. Ontological modularity and uncertainties guide and support combinations and extensions of the developed spatial ontologies. The modules' applicability and adequacy are evaluated individually and in different spatial application scenarios.

Zusammenfassung

Die vorliegende Dissertation mit dem Thema "Modular Ontologies for Spatial Information" beschäftigt sich mit der Entwicklung und Anwendung von räumlichen Ontologien in verschiedenen Anwendungsszenarien und untersucht die dafür notwendigen Methoden und Techniken.

Dabei wird zunächst untersucht, auf welche Arten räumliches Wissen modelliert werden kann. So wird analysiert, welche Typen und Konzeptualisierungen von räumlichen Informationen konkret unterschieden werden können. Die Arbeit beschäftigt sich daher zunächst mit thematisch unterschiedlichen "Perspektiven", die unterschiedliche Beschreibungen räumlicher Informationen bieten. Beispielsweise können quantitative, qualitative, abstrakte, domänen-spezifische oder multimodale Beschreibungen räumlicher Informationen unterschieden werden. Diese unterschiedlichen Perspektiven werden daran anschließend in sogenannten Ontologien formalisiert. Ontologien sind logische Spezifikationen, die semantische Kategorien modellieren und klassifizieren können. Technisch wird hierfür insbesondere OWL und Protégé sowie CASL und HeTS eingesetzt.

Zusätzlich wird in dieser Doktorarbeit untersucht, inwieweit die entwickelten Ontologien sowohl inhaltlich als auch technisch verknüpft werden können. Im Mittelpunkt steht hierbei die Modularität von Ontologien, denn sie ermöglicht es einzelne Ontologiemodule auf unterschiedliche Weisen miteinander zu kombinieren. Es werden drei Formen unterschieden: Extension/Refinement, Matching und Connection. Diese werden miteinander verglichen und abhängig von thematisch-inhaltlichen Kombinationszwecken für unterschiedliche Anwendungsanforderungen eingesetzt und für konkrete Ontologiemodul-Kombinationen angewendet.

Ein weiterer Untersuchungsgegenstand ist die Integration unsicheren Wissens in die Ontologiemodule. Beim Einsatz von den entwickelten Ontologien in konkreten Anwendungen sind teilweise nur unsichere, vage, oder mehrdeutige Informationen vorhanden. Bisher existieren hierfür keine standardisierten Methoden, wie logisch-strikte Ontologien mit unsicherem Wissen in Verbindung gebracht werden können. Daher wird in dieser Dissertation untersucht, wie diese Verbindung technisch und inhaltlich realisiert werden und in verschiedenen Anwendungsszenarien zum Einsatz kommen kann.

Die in der Arbeit entwickelten Ontologiemodule für räumliche Informationen, deren Kombinationsmöglichkeiten und deren Verknüpfung mit unsicherem Wissen kommen schließlich in drei Anwendungen zum Einsatz und werden hier evaluiert. Im Bereich Architekturdesign und Assisted Living werden die räumlichen Ontologiemodule eingesetzt, um Designanforderungen und Assistenzautomatisierung zu modellieren und zu prüfen. Im Bereich Bildverarbeitung werden die räumlichen Ontologiemodule eingesetzt, um Objekt- und Szenenklassifikation zu unterstützen. Im Bereich Sprachverarbeitung werden die räumlichen Ontologiemodule eingesetzt, um räumliche Sprache mittels qualitativer räumlicher Repräsentation zu interpretieren.

Das Ergebnis dieser Doktorarbeit zeigt, dass unterschiedliche räumliche Perspektiven durch formale Ontologien modelliert werden können, indem diese Ontologien modular entwickelt und hinsichtlich unterschiedlicher inhaltlicher und technischer Aspekte kombiniert werden. Zusätzlich wird unsicheres Wissen mit diesen modularen Ontologien verknüpft, um notwendige Unsicherheiten in Anwendungsszenarien abzudecken. Die Gesamtmodellierung der resultierenden Ontologien mit den entsprechenden Erweiterungen werden abschließend in verschiedenen Anwendungsszenarien erfolgreich zum Einsatz gebracht und evaluiert.

Acknowledgements

I would like to thank foremost my PhD supervisors John Bateman and Kerstin Schill. They supported and encouraged my work on this thesis and gave me the opportunity to pursue my research topics. I am also grateful for having been a part of the SFB/TR8 Spatial Cognition Research Center, subproject I1-[OntoSpace], at the university of Bremen. Without this research environment and its administrative support several collaborations and projects would not have been possible.

I also owe huge thanks to many colleagues and friends, who helped me through this long, exhausting, and sometimes what seemed to be never-ending journey. They made the past years fun and interesting, and I was able to learn a lot from them. Among them, in random order and without a claim of completeness, are: Emma, Oliver, Linn, Mehul, Parisa, Carl, Thora, Carsten, Christian, Rob, Falko, Shi, Immanuel, Yohei, Till, Thomas, Niclas, and Dimitra.

I would like to thank my family, Sigrid and Jessica, for their continuous support and encouragement.

Finally, I am deeply indebted to Mathis, who has not only been my unofficial third reviewer but also inspired me at every stage of this thesis.

Thank you all so much, guys!

Contents

CONTENTS

List of Figures

List of Tables

Listings

1

Introduction

For the representation of spatial information a system needs to define which aspects of space should be represented and how the aspects are specified. For example, its requirements could be to represent spatial objects (cubes, cones, squares, circles), spatial relations (right, left, straight, diagonal), outer space (planets, stars, rotations, paths), or geographical types (islands, mountains, countries, capitals). This raises the questions: how many types of spatial information exist that can be represented by the system; what are the characteristics of these types; and how are the different types related with each other. More specifically: is there a way to generally categorize these types and their properties? Exactly these questions are addressed and analyzed in this thesis.

The research literature shows that space can be described, analyzed, conceptualized, and structured in various ways. Several approaches have been developed to describe spatial information with different, individual characteristics. Among them are formal models of space (e.g., spatial calculi [Cohn and Hazarika, 2001]), mathematical and geometrical constructs (e.g., co-ordinate systems [Tarski, 1959]), cognitively perceived or conceptual space (e.g., proximity, separation, order, enclosure, continuity [Piaget and Inhelder, 1997]), imaginative recurring patterns (e.g., image schemata [Johnson, 1987]), spatial objects (e.g., object level and form level [Olson and Bialystok, 1983]), application-specific spatial information (e.g., geographical information systems [Frank, 2001]), or spatial relationships expressed in natural language (e.g., locative prepositions [Herskovits, 1986]), just to name some and to illustrate the large variety.

The aim of this thesis is to generally distinguish and classify different types of spatial information in order to make their characteristics explicit. On the basis of this distinction, ontological specifications are developed that reflect, relate, and combine the different types of spatial information. The resulting ontological representations are applied, extended, and evaluated in spatial systems that represent spatial information for their tasks or purposes. The collection of these ontological representations reflects the illustrated variety of spatial information necessary to understand meanings of space.

The diversity of spatial types of information is thus distinguished and specified in this thesis. We show that individual spatial models and disciplines provide different perspectives, in particular, we will differentiate quantitative, qualitative, abstract, domain-specific, and modal perspectives. In this thesis, these perspectives are defined, analyzed, and categorized to reflect and structure the diversity of spatial representations. For example, their implementations can range over the conceptualization of space in natural language, perceptual categorizations of space by means of visual features, or application-based classifications of space in architectural design. Furthermore, the adequacy and applicability of this approach, its results, and potential extensions are evaluated by comparing the ontological representations with the spatial domain and by applying the ontological representations in spatial systems.

These goals are methodologically achieved by using techniques from formal ontology and ontological engineering. In particular, modularity methods are analyzed and applied on the basis of their thematic and technical applicability. Also, uncertainty theories are selected in application-specific scenarios to support individual tasks and purposes that have to deal with uncertain information. To establish the distinction of the spatial perspectives and to analyze the variety of spatial information, ontological and formal models of space, empirical findings and experimental results from the area of spatial cognition, and insights from different disciplines related to space are taken into account.

In summary, this thesis investigates different types of spatial information and it provides perspectival specifications for them by using modular ontologies. These modules reflect how space can be perceived, described, interpreted, used, and formalized from different perspectives. They are also developed to support various application-specific purposes, aims, and tasks. Their applicability and usefulness are illustrated and evaluated in these spatial application scenarios.

1.1 Relevance

The main research contribution of this thesis is the ontological analysis of the variety of spatial information. In particular, we investigate how ontological spatial representations have to be structured and specified to adequately model this variety. Necessary extensions of these ontological representations are examined, and the ontological specifications are applied and evaluated in spatial systems.

From a theoretical point of view, the variety and diversity of spatial information is analyzed from multiple perspectives and different disciplines. Individual characteristics of space are differentiated and their contents and spatial aspects are analyzed and compared. In particular, heterogeneous spatial types are classified and their differences are made explicit. Nevertheless, as these different types may refer to similar spatial information, they may formalize overlapping or complementing spatial descriptions, which are related or combined with each other. Hence, this thesis presents formal representations of thematically distinguishable spatial definitions, and it provides methods and tools to develop, evaluate, and apply these representations as ontological modules.

From an application-specific point of view, spatial assistance systems or spatial agents need some spatial representation that provides them with a knowledge structure for their environment or tasks. Particularly, when cognitively-motivated spatial systems are supposed to interact with humans and their environment in a human-like way. In order to achieve this behavior, such systems need to be able to perceive and classify their surroundings. In addition to the representation of perceived environmental entities, however, systems also have to be endowed with a representation of more abstract entities and relationships, which are not directly perceivable but which support systems to derive further information or categorize existing information. Again, these different types of representations are distinguished and specified as ontological modules that can be combined or extended based on application-specific requirements.

For these purposes, this thesis addresses how different types of spatial information can be specified by ontology modules. It analyzes formal and technical requirements to reflect the diversity of spatial information and it examines the aspects ontological specifications have to satisfy for the different perspectives on space. The result is a perspectival ontological model to represent these different perspectives. It classifies the perspectives and guides the development of new spatial ontology modules. Several application examples demonstrate and evaluate the implementation and usability of the model, and they illustrate how the various distinctions of spatial information can support application-specific tasks and domain-specific requirements.

Hence, the aim of this work is to establish an ontological formalization for spatial information from different perspectives. This formalization is directly supported by ontological modularity techniques, and it provides extensions for combinations and aspects of uncertainty. The resulting ontological formalizations give a clear distinction of spatial perspectives, which allows a separation into thematically different modules that structure the spatial domain. Dependencies and connections across modules are defined and their use is evaluated in spatial application contexts.

1.2 Motivation

Specifications of spatial ontology modules are supposed to thoroughly define the domain of space, provide a formal structure and categorization for spatial entities, and help spatial systems to classify their spatial data. Especially as space is inherently involved in almost any domain, its ontological specification can support a broad range of tasks and systems. For example, spatial assistance systems need to understand abstract spatial entities, domain-specific aspects and features of spatial entities, dependencies between specific types of entities, context-sensitive interpretations of information, intentions of communication partners, causes and effects of actions, or possible changes in their environments. These systems may use spatial knowledge structures that are based on ontological specifications of space covering the variety of those different types of spatial information, thus providing them with information on spatial entities and relations they need to know.

This thesis primarily aims at investigating the way ontological formalizations can represent the diversity of spatial information. In particular, the result should give a clear indication how the diversity can be classified ontologically and how making this explicit can help to develop and use spatial ontology modules. The distinction of spatial perspectives and their ontology modules is also intended to serve as a basic understanding and structured categorization of the different types of spatial information. It should thus be feasible to determine which spatial types exist in a specific context, how these types can be grouped and structured, and how they can be specified, grounded, and applied. Users and systems should then be able to select spatial ontology modules for concrete applications or specific tasks by analyzing their application contexts with regard to the different spatial perspectives.

The topics investigated in this thesis contribute to the research areas that particularly focus on the heterogeneity of spatial information and its diverse interpretations. These areas spread over the disciplines formal ontology, spatial cognition, and spatial information, where tools from ontological engineering are used to model the various aspects of different spatial types in a multifaceted way [Bateman et al., 2010a; Bennett and Agarwal, 2007; Takeda et al., 1995; Timpf et al., 1992].

1.3 Methodology

Ontologies are formal representations that describe and classify meanings of categories. They "can be thought of as establishing common agreements among numerous experts on the logical meaning of certain terms in a particular field" [Kutz et al., 2010b]. For the spatial domain, ontologies can provide a standardized representation for space as a terminological common ground; they can be applied in spatial assistance systems, and used to exchange and integrate information across sources. This thesis analyzes how spatial ontological specifications have to be modeled to describe the designated spatial characteristics, which of these characteristics need to be taken into account, which requirements spatial ontologies have to satisfy, and how spatial ontologies can capture the diversity of spatial information in order to reflect heterogeneous representations of space.

The work is based on methods and best practices in ontological engineering. The thematic analysis of the domain as well as the design and implementation of the spatial ontology modules and their evaluation are guided by principles from the field of formal ontology (cf. section 2.3.2). This primarily results in an emphasis on a close connection between the ontological specification and the actually modeled domain. The intention in formal ontology is to adequately and precisely model the thematic characteristics through ontological representations as close as possible to reality. The spatial ontology modules developed in this thesis are consequently based on external resources, spatial formats, existing models, or empirical findings to achieve this adequate representation. This also causes the multidisciplinarity of developing spatial ontology modules,

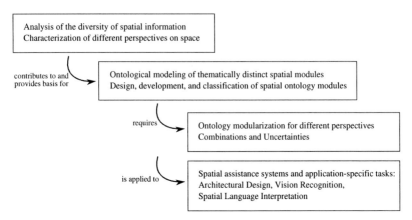

Figure 1.1: Thesis Overview. Main goals and aspects to be analyzed and achieved in this thesis.

as insights from spatial cognition, spatial mathematical models, spatial domain-specific systems, and general ontological research can contribute to the development.

Application scenarios of spatially-aware systems demonstrate how the spatial ontologies are applied, how they can support tasks of spatial systems, and how they can be used in practice. The evaluation shows to what extent the spatial systems can benefit from the distinction of different perspectives on space. It also shows where extensions and combinations of the spatial modules are necessary. The spatial systems analyze and demonstrate usability, applicability, and adequacy of the developed spatial ontology modules. Methods from ontology evaluation, a research area of its own, are also introduced and employed where appropriate. In particular, empirical data-driven analyses and statistics are used to evaluate the design and task performance of the developed ontology modules.

Figure 1.1 illustrates the essential investigations of this thesis. A general analysis of the diversity of spatial types of information results in ontological representations of these spatial types, namely the spatial perspectives. For developing ontologies for these perspectives, necessary requirements are examined, namely modularizations, combinations, and uncertainties. The ontologies are then applied and evaluated in specific spatial systems.

1.4   Overview of Chapters

This chapter has laid out the goals, purposes, and methods of the thesis. The next chapter *"Survey of Ontologies and Spatial Information"* summarizes related work in the field of spatial information. Ontologies and their development process are presented, and existing specifications for the spatial domain are discussed. Recent trends in the modularization of ontologies are summarized in order to indicate its importance and anchoring in the field of computer science. Aspects of uncertainties are introduced as a new direction in ontological engineering, and their impact on ontologies for space is illustrated. Finally, spatial cognitive aspects and their influence on ontological specifications for space are outlined.

In chapter 3 *"Types of Space and Spatial Information"*, the different types of spatial information are presented. This includes the ways in which they can be categorized and structured. After introducing different perspectives for describing aspects of space, their relations and combinations are identified. The result of chapter 3 is a model for different spatial perspectives that contributes to an overall specification of the spatial domain by pointing out characteristics, differences, and combinations.

In chapter 4 *"Modular Ontologies for Spatial Information"*, spatial ontology modules are developed on the basis of the spatial perspectives that reflect the modularly structured and heterogeneous types of spatial information. Individual thematically different ontologies are introduced and explained with regard to their design, modeling decisions, purposes, requirements, evaluations, and applicabilities. Technical details and formal analyses of the ontologies as well as their reasoning practicability are discussed.

Chapter 5 *"Combinations and Uncertainties of Ontology Modules"* enriches and extends the ontology modules introduced in the previous chapter with respect to particular types of combination techniques and aspects of uncertainties. Relating the thematically different ontologies with each other requires the formalization of their connections. Requirements and solutions for such combinations are thus described. The influence of uncertainties not only for combination aspects but also for ontological specifications in general is also presented in chapter 5.

In chapter 6 *"Application and Evaluation"*, spatial application scenarios are presented that use and integrate the developed spatial ontology modules. These applications demonstrate how the perspectival ontological specifications for space can be applied and used in different contexts. In particular, applications in the fields of assisted living, visual recognition, and spatial language interpretation are presented. Practicability, adequacy, and usability of the spatial ontology modules are also evaluated here.

Finally, the results and achievements of this thesis are summarized and discussed in chapter 7 *"Conclusions and Outlook"*. Related aspects and further extensions that could be developed for the spatial ontology modules are pointed out. Future directions, open questions, and further research possibilities close the chapter.

2

Survey of Ontologies and Spatial Information

This chapter gives an overview of the research topics ontologies and spatial information. As this thesis aims at formalizing spatial information by specifying ontologies for it, both topics are relevant and thus surveyed in the following.

First, the field of ontologies and ontological engineering in the area of knowledge representation and reasoning is presented. The section starts with a short historical overview followed by details of technical aspects, development processes, and evaluation strategies from both theory and application. Furthermore, aspects of modularity and uncertainty in the area of ontologies are presented, and their relevance for the ontological specifications presented in chapters 4 and 5 is shown.

Second, the field of space, spatial information, and spatial cognition is introduced. Different types, aspects, and purposes of spatial information are surveyed together with the research communities involved in this topic, reflecting the heterogeneous nature of spatial information. Analyses and treatments of space and spatial applications in cognitive science, artificial intelligence, and related disciplines are shown.

2.1 Knowledge Representation and Reasoning: Historical Background

Early work in *knowledge representation and reasoning* (KR) has its roots in the 1970s, when the idea of *physical symbol systems* was introduced by Newell and Simon [1976]. In their paper, the authors describe the representation of knowledge by using symbols and reasoning with knowledge by manipulating these symbols. This form of KR is supposed to resemble the processing of information by humans and thus resemble "intelligent behavior" [Newell and Simon, 1976] in the sense of artificial intelligence (AI). The symbols used refer to objects or expressions that form the knowledge base of an agent. Symbol tokens are expressed by using formal logic (representation) and logical operators are used to provide manipulation of symbols for information processing (reasoning).

KR was further promoted by its use in expert systems in the 1970s. The MYCIN system is a well-known medical example of this [Buchanan and Shortliffe, 1984]. It formulated medical parameters as *data triples*, i.e., (1) objects had (2) attributes that were assigned (3) values. Together with a rule and inference engine and probabilities, this structured data was used for medical diagnoses. In particular, the data was supposed to reflect the expert knowledge of doctors or medical staff [Buchanan and Shortliffe, 1984, p. 153]. The aim of the structured data was to detect relations and dependencies, to get a better understanding of the overall data. This particular goal is said to be reached by the MYCIN system by providing a highly flexible and intelligible knowledge base [Buchanan and Shortliffe, 1984, p. 671] with regard to the way it defined its facts.

In the 1980s, the next cornerstone in KR was the development of Cyc, a knowledge base that is still in use and further development today.[1] An example of elements defined in the Cyc knowledge representation is shown in figure 2.1. Cyc's aim is to specify all encyclopedic and common-sense facts about the world and to be able to draw inferences over these facts. Cyc's data structure is again based on triples and a LISP-like notation is used for formulating complex expressions [Lenat and Guha, 1990]. Cyc currently defines more than 4.6 million such expressions and, as a consequence, the knowledge base has been re-structured into microtheories, which are hierarchically connected [Taylor et al., 2007]. These microtheories range from high-level theories with broad information (e.g., about time and space) to fine-grained theories with specific context information (e.g., about US presidents).

With the development of KR systems, the need for formalizing and structuring knowledge base data increased. The use and development of *semantic networks* (SNs) was one major result of this, although SNs are based on ideas first presented in the 1950s [Richens, 1956]. The structure of SNs consists of nodes and links between nodes,

---

[1]http://www.cyc.com (visited on July 05, 2010)

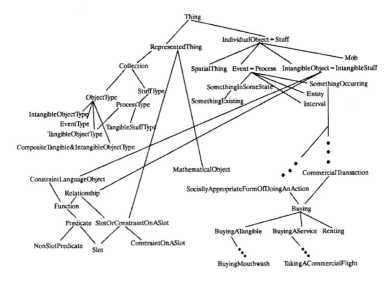

Figure 2.1: Cyc Example. Some predicates defined in the version of Cyc presented in the primary Cyc publication by Guha and Lenat [1990].

i.e., a graphic notation with nodes and arcs is used for representing knowledge [Sowa, 1992]. In contrast to triples as used by the MYCIN system, objects are not just assigned values via so-called *slots*, but objects are related also to other objects via so-called *links*, i.e., nodes mostly represent objects or events and the arcs represent links or relations between them. In general, semantic networks are used as an informal knowledge representation schema. Some of them have, however, been specified in a formally defined system of logic, and different kinds of such systems can be distinguished [Sowa, 1992], e.g., in the case that links are only used to reflect subtype or implication relationships between nodes, or in the case that links can not only be true but also marked with a modal operator. Semantic networks allow direct access to information about certain facts in the knowledge base, for instance, facts about an object 'President' are available by collecting all of its follow-up links [Allen and Frisch, 1982]. Research in SNs has been carried out in a large variety of disciplines including cognitive psychology, artificial intelligence, lexicography, logics, philosophy, psycholinguistics, linguistics, computer science, and related interdisciplinary disciplines [Helbig, 2006, p. 16]. SNs have also

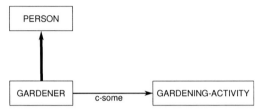

Figure 2.2: KL-ONE Example. The figure illustrates the Gardener definition adapted from Woods and Schmolze [1992], specified in the KL-ONE formulation on page 13.

been specified in logic-based languages, e.g., formulated with lambda notation [Woods, 1975].

The next milestone in KR was the advancement of *frame languages* as a knowledge representation framework, first introduced by Minsky [1974] on the basis of psychological results. Although semantic networks are already able to collect relevant information about one object by using the object's links, frames provide an even stronger focus on the objects as such. One frame reflects information about one object or *category*. In particular, it "is a data-structure for representing a stereotyped situation" [Minsky, 1974]. A frame is filled with *slots*, which can either point to values, other frames, or methods of the frame (in contrast to SNs). As an extension to SNs, frames make a clearer distinction between the objects themselves and the frames that provide a generic 'template' for objects, i.e., a frame provides prototypical or possible slots that an object of that frame can have. Frames can further be related hierarchically with each other, i.e., frames can inherit their slots from super-frames, and there can exist one root frame that subsumes all other categories (frames). It has been shown that frame languages can be formulated as fragments of first-order logic, thus their expressiveness is comparable to first-order logic [Hayes, 1980]. They provide, however, a strongly object-oriented modeling method, which led to the idea that more efficient reasoning techniques could be provided by using a hierarchical structure and by using decidable fragments of first-order logic, resulting in the development of *description logics*.

Based on this connection between frame languages and logics [Hayes, 1980], description languages then provided clearly defined, logic-based, formal semantics while aiming at efficient reasoning strategies. An example is the description language KL-ONE [Brachman and Schmolze, 1985], which provides an object-centered knowledge representation that uses structural subsumption algorithms. An example of specifying an object Gardener is illustrated visually in figure 2.2. The following KL-ONE specification (taken from Woods and Schmolze [1992]) defines that a Gardener is subsumed by a Person with at least one hobby that has to entail a Gardening-Activity:

Definition. *(cdef GARDENER (and PERSON (c-some Hobby GARDENING-ACTIVITY)))*

The description logics available nowadays are a family of logics that are decidable fragments of first-order logics, which take into account an object-oriented view and provide efficient reasoning techniques by using tableau-based algorithms, automata-based or resolution-based approaches [Baader et al., 2007]. Up to now, description logics are the primary standard for decidable KR systems. For this reason, they are used as a specification language for the ontologies developed in this thesis. They are introduced in more detail in section 2.3.1.

As a branch of KR, the field of *knowledge engineering* was established in the 1980s. Knowledge engineering primarily investigates and models processes for machine-encoding expert knowledge into computer systems. This encoded knowledge does not necessarily have to comply with an expert's cognitive model but with an adequate representation of it to infer further knowledge. The resulting knowledge base can then be used in problem-solving tasks. In knowledge engineering, building the knowledge base is often divided into different steps in the development phase, such as knowledge elicitation, design, formalization, interpretation, and application [Studer et al., 1998]. The main emphasis lies on a thorough analysis of the exact modeling strategies and decisions for appropriately reflecting the modeled knowledge (see also section 2.3). Along with knowledge engineering, tool support particularly for developing and reasoning with knowledge representations has been developed [Studer et al., 1998].

As a consequence, the research topic *ontology* emerged as a specific form of symbolic representation, overlapping with research in description logics, knowledge engineering, and formal modeling. It builds on design criteria supposed to reflect conceptual knowledge, it brings together symbols (syntax) and their semantics (model-theoretic semantics), and it provides applicable reasoning techniques. In particular, two complementary aspects, one from formal ontological research and one from ontological engineering, contribute to this research field, as introduced in detail in the next section.

Even though *Nouvelle AI* introduced the *connectionist paradigm* in the late 1980s, which uses neural information processing and network-learning for reproducing "intelligent behavior" [Brooks, 1999] and which is inspired by the idea that symbols are ignorable and knowledge is supposed to be built by low-level or sensorimotor data and interaction, the area of formal logics continues to be a major branch of KR for symbolic knowledge representation. Although connectionist and symbolic approaches are predominantly investigated separately, recent directions now focus on combinations of both neural and symbolic methods for the support of intelligent behavior and for realizing synergies between both methods [Hammer and Hitzler, 2007]. Such *neuro-symbolic* approaches are especially relevant to application-oriented knowledge representations, as they are related to low-level or sensorimotor data. Hence, neuro-symbolic combinations are also relevant to the ontologies developed and applied in this thesis. In particular, applications often have to deal with different kinds of *uncertainties* caused by sensorimotor data. Thus, an outline of uncertainties is presented, and combinations specifically of uncertainties and ontologies are then discussed in section 5.2.

Nowadays, the field of KR continues its investigation of how to use logical symbols to reflect information in the most appropriate way. The goal of the discipline can be summarized as follows : "Knowledge representation and reasoning is the area of Artificial Intelligence (AI) concerned with how knowledge can be represented symbolically and manipulated in an automated way by reasoning programs." [Brachman and Levesque, 2004, p. xvii]

2.2 Knowledge Representation and Reasoning: Ontological Engineering and Formal Ontology

The term ontology as a particular type of knowledge representation was coined in the 1990s. Its definition varies slightly in the different disciplines, i.e., different communities in the field of computer science or information science have established their own definitions about the meaning of the term *ontology*. One general definition is that "ontologies are content theories about the sorts of objects, properties of objects, and relations between objects that are possible in a specified domain of knowledge. They provide potential terms for describing our knowledge about the domain" [Chandrasekaran et al., 1999]. This definition points out that ontologies consist of objects and their properties and relations, i.e., ontologies have a strong object-oriented view towards a domain. Objects, properties, and relations are also the elementary components an ontology can specify. The definition also points out that an ontology describes one particular domain, i.e., usually a selected part of general world knowledge.

In the area of ontology engineering, a more specific definition is given by Gruber: "An ontology is an explicit specification of a conceptualization" [Gruber, 1993]. This definition emphasizes that an ontology makes explicit what it describes, namely the concepts of a certain domain. An extended version of this definition is that "an ontology is a formal, explicit specification of a shared conceptualisation" [Studer et al., 1998]. Here, the emphasis lies on *formal*, i.e., the ontology is formalized in a logic-based language, and on *shared* among different parties, i.e., either a community has agreed on a certain specification of the concepts or a consortium has established a standardized specification to which a community adheres. In the same tradition, "a shared understanding of some domain of interest" [Uschold and Gruninger, 1996] emphasizes that an ontology specifies a particular restricted part of a domain. In many cases, these definitions are associated with certain development processes and concept specification guidelines (see section 2.3).

In the context of the *Semantic Web* [Berners-Lee et al., 2001], an approach that aims at semantically enriched access to web information by making Internet web services inter-operable based on ontology documents, ontologies are mainly seen as a knowledge infrastructure for distributed and web-based intelligent agents (applications), namely semantic web services. Hence, "an ontology is a document or file that formally defines the relations among terms" [Berners-Lee et al., 2001]. In a similar direction, Hendler defines an ontology as "a set of knowledge terms, including the vocabulary, the semantic interconnections, and some simple rules of inference and logic for some particular topic" [Hendler, 2001]. Thus, an ontology provides a terminology (the vocabulary) of a domain, relationships within and across domains, and some inherent inference methods supported by the logic, in which the ontology is formulated. In the field of the Semantic Web, ontologies are oftentimes simply referred to as documents.

In the field of database systems, ontologies are seen as an extension to the database schema. While databases are intended to deal with large amounts of data, ontologies are supposed to provide an enriched vocabulary and further inference techniques to databases. "In the context of database systems, ontology can be viewed as a level of abstraction of data models, analogous to hierarchical and relational models, but intended for modeling knowledge about individuals, their attributes, and their relationships to other individuals" [Gruber, 2009]. Ontologies are used in this area particularly to support interoperability between different databases containing heterogeneous types of information. Heterogeneous interoperability is also an essential aspect of the ontologies developed in this thesis (see section 5.1).

In information science, the definitions of ontologies are motivated less technically than those above. Here, ontologies are mostly seen as artifacts that may even take into account modeling recommendations from philosophy. This particular research discipline is nowadays called *formal ontology* and has contributions primarily from computer science and philosophy. Guarino defines an ontology as "an engineering artifact, constituted by a specific vocabulary used to describe a certain reality, plus a set of explicit assumptions regarding the intended meaning of the vocabulary words" [Guarino, 1998]. Hence, it is most central that an ontology describes some 'real' facts. It is a human-made product that determines the semantics of this domain and that makes its rationale explicit. The domain is defined by using certain structures (objects, properties, and relations) and axiomatizations:

> In effect, the term 'ontology' applies to virtually any structure resembling, to some extent, a set of terms hierarchically organized which may be put in a machine-processable format. [. . .it] may include: (1) a set of terms (classes, categories, concepts, words), (2) an axiomatic theory or a set of propositions, (3) the content [. . .] of a knowledge base in general [. . .] Indeed, some even speak of the task of building an ontology as a matter of conceptual modeling or conceptual representation (a conceptual system would be a model of reality). [Grenon, 2008, p. 29 f.]

In summary, ontologies are defined, in engineering, as logical tools for the knowledge representation of some domain; in the Semantic Web, ontologies are defined as documents for distributed agents; in the database community, ontologies are defined as an interoperable schema for heterogeneous data; and in formal ontology, ontologies are defined as an engineering artifact of some real world domain. Although all these ontology definitions share a common ground in the different disciplines, their variations are subtle. For instance, in the Semantic Web and similarly in the database community, ontologies are merely a technical instrument possibly exchangeable with other formalisms. There exists a large number of such ontologies with large taxonomies and few axiomatizations. In engineering, ontologies are primarily a method to reflect the knowledge structure of a certain domain, which has to be applicable to some AI systems. In formal ontology, every ontology has to be carefully specified according to conceptual representations for parts of a real domain, which may even result in specifying only one foundational ontology reflecting all real domains.

In general, however, the ontology definitions from the different disciplines have in common that they describe a certain domain by specifying classes or elements, their properties and their relations, represented by a hierarchical representation that reflects a subsumption relation between classes. They are all formulated in some logical language. Also, they logically consist of unary (categories) and often binary or in general n-ary (relations) predicates and their axioms. The hierarchical organization of the categories is called a *taxonomy*. More technically, an ontology provides an overview of existing entities and relations in the domain, which can be used for classification, query answering, consistency analysis, and dependency identification. A distinction is also made between the formalization and the instantiation: the formalization describes what types of classes and relations can exist in the domain, whereas an instantiation (if defined) describes a particular state or situation that has to satisfy the axioms given by the formalization.[2]

Across disciplines, the aim of an ontology is to provide a *controlled vocabulary* that contains all terms that can possibly occur in a domain. Hence, an ontology enables an agent or system to know the symbols it can use. This vocabulary consists only of terms with unambiguous meanings offering a common understanding of those terms [Chandrasekaran et al., 1999]. As ontologies precisely characterize the different types of concepts or categories and their relations for a given domain, they determine the underlying organization or the *ontological commitment* [Guarino et al., 1994], i.e., they confirm the meanings of the categories as specified. Users and systems can thus comply with this meaning using the same ontology. The major reason and purpose to develop and apply ontologies is their possibility to share and reuse knowledge.

---

[2]The distinction between the abstract formalization of the domain and concrete instantiations are, for example, supported by description logics, introduced in section 2.3.1.

The ontologies developed in this thesis are primarily rooted in the discipline of formal ontology.[3] They commit to a specific way of viewing entities and relations, which is also called the *conceptualization* in information systems [Olivé, 2007, p. 11]. The ontologies developed in this thesis are supposed to reflect a conceptual model of parts of a real-world domain, which is space and different types of spatial information. Hence, the aim is to reflect space as closely as possible as it is perceived by humans or as it is used (and usable) by spatial assistive applications. The aim is to make these ontologies usable for spatial applications or tasks, as they precisely reflect actual spatial information. The aim is also that the resulting ontologies contribute to the research field of spatial information and spatial cognition (see section 2.6) by providing an overview of the different aspects of space that occur in reality. As the ontologies are, however, intended to be used within spatial systems, they also need to provide technical applicability and usability. Thus, they are mostly formulated in decidable logics and written in standardized data formats thereof (see section 4.1.2).

2.3 Ontology Languages, Design, and Development

This section gives an overview of technical and methodological aspects in ontology design and development. These aspects have to be considered when developing new ontologies or revising ontologies and are thus relevant for the ontologies developed in this thesis. The following section introduces logical formalisms used to specify ontologies as well as methodological techniques used to capture knowledge represented in ontologies. Tools that support the formalization and development process are referenced accordingly.

---

[3]Even though the term "ontology" is used in this thesis in the field of computer science, it originally stems from the field of philosophy and was first introduced in Aristotle's Metaphysics, in which ontology was the discipline that describes the things that exist [cf. Loux, 2002, p. 3]. Regardless of goodness or truth, things are described by their general features. A basic distinction between things is based on generalization and specialization, i.e., which things are more general or more specific than others, a distinction that is reflected by the subsumption hierarchy of ontologies also in the definitions from computer and information science. Although ontologies in this thesis are not based on philosophical considerations, some of the considerations about general concept distinctions are utilized in section 4.1.

2.3.1 Logical Formalization

The specification of a knowledge base is commonly formulated in a logic-based language. A knowledge representation language consists of its *syntax*, i.e., a set of symbols and a way of formulating well-formed sentences given the symbols, its *semantics*, i.e., the meaning of expressions that are formulated by the syntax, and its *pragmatics*, i.e., the use of meaningful expressions. According to Brachman and Levesque [2004], meanings of expressions are captured by an interpretation $\Im$, which is a pair $\langle \mathcal{D}, \mathcal{I} \rangle$ with $\mathcal{D}$ as the domain (a nonempty set of symbols) and $\mathcal{I}$ as the interpretation mapping. The interpretation determines the truth value and satisfiability of sentences. In particular, if α is a formula and $\Im \models \alpha$ we say that α is true in the interpretation $\Im$. If S is a set of sentences and $\Im \models S$ then $\Im$ is a *logical model* of S.

For example, a domain $\mathcal{D}_1$ may consist of the symbols City, Country, River as unary predicates, located_in, flows_through as binary predicates, and Bremen, Germany, Weser as named individuals. Facts of the domain can then be formulated using logical quantifiers and connectives to specify complex expressions. In the example domain $\mathcal{D}_1$, every city is supposed to be located in a country; this is expressed by the following formula:

$$\forall x.\mathsf{City}(x) \Rightarrow \exists y.\mathsf{located\_in}(x, y) \land \mathsf{Country}(y)$$

Types of the individuals in the domain can be defined by using the unary predicates, such as River(Weser) or Country(Germany).

In this thesis, an ontology is defined as follows:

Definition (Ontology). *An ontology O is a tuple $\langle C, R, I, A \rangle$, for which C is the set of concepts, R is the set of relations between concepts, I is the set of instances (also called individuals) that instantiate the concepts and relations, and A is the set of axioms or constraints that use concept, relation, or instance names and logical operators. The sets C, R, I, and A are pairwise disjoint.*[4]

Although logical expressions can be used to build a knowledge base in general, ontologies aim at taking into account common-sense information concerning the domain and formal structuring aspects as defined above. In particular, objects naturally fall into categories; they are members of different categories, there are more general and more specific categories, objects have parts and certain relations to other objects, which

---

[4]I can be empty, and some ontologies even require I to be empty (cf. section 4.2.3.1).

leads to an object-oriented approach for formalizing a domain. As formulating this information mostly requires complex and compound predicates and expressions in order to specify varieties of categories and their relationships, the ontology language primarily defines *descriptions*. Ontologies therefore are often formulated in *description logics* (DL), a family of decidable logics [Baader et al., 2003].

"Description logics are a family of knowledge representation languages that can be used to represent the knowledge of an application domain in a structured and formally well-understood way" [Baader et al., 2007]. They make a distinction between termi-nological and assertional information: terminological information (TBox) defines the categories of a domain and their constraints and assertional information (ABox) defines the facts or a specific model according to definitions of the TBox. The TBox contains *categories* (also called *concepts* or *classes*), i.e., unary predicates, and *relations* (also called *roles*), i.e., binary predicates. The ABox contains *instances* (also called *individu-als*), i.e., constants, and their properties and relations. Regarding the example domain above, City, Country, River are categories in the TBox, located_in, flows_through are rela-tions in the TBox, and Bremen, Germany, Weser are instances in the ABox. Furthermore *constraints*, i.e., logical sentences, about terminological information can be formulated in the TBox. For example, that all cities are located in countries is expressed by

$$\text{City} \sqsubseteq \exists \text{ located\_in.Country}$$

i.e., all instances that are a City are necessarily located_in a Country. Formulating that cities can only be located in exactly one country can be done by using cardinality restrictions, expressed by, e.g.,

$$\text{City} \sqsubseteq \; = 1 \text{ located\_in.Country}$$

A TBox is a finite set of such formulas, which are also called *general concept in-clusions* (GCI). A set of GCIs specifies an ontology based on a logic-based semantics. Table 2.1 shows an overview of the different constructs available in the DL $\mathcal{ALC}$. This description logic is the smallest one that is propositionally closed, i.e., it provides (implicitly or explicitly) intersection, union, and negation of class descriptions. The semantics are given by the set-theoretic interpretation function $\mathfrak{I} = (\Delta^{\mathfrak{I}}, \cdot^{\mathfrak{I}})$, which reflects the interpretation $\langle \mathcal{D}, \mathcal{I} \rangle$ (see above), C and D are category and R relation

Table 2.1: **DL Syntax for $\mathcal{ALC}$.** DL constructors, their syntax and semantics with examples for the logic $\mathcal{ALC}$ (cf. Baader et al. [2007]).

DL Constructor	Syntax	Semantics	Example
atomic concept	A	$A^{\mathfrak{I}} \subseteq \Delta^{\mathfrak{I}}$	River
atomic role	R	$R^{\mathfrak{I}} \subseteq \Delta^{\mathfrak{I}} \times \Delta^{\mathfrak{I}}$	flows_through
conjunction	$C \sqcap D$	$C^{\mathfrak{I}} \cap D^{\mathfrak{I}}$	River $\sqcap$ SalineWater
disjunction	$C \sqcup D$	$C^{\mathfrak{I}} \cup D^{\mathfrak{I}}$	River $\sqcup$ Stream
negation	$\neg C$	$\Delta^{\mathfrak{I}} \setminus C$	$\neg$ Stream
existential restriction	$\exists R.C$	$\{x \mid \exists y.\langle x,y \rangle \in R^{\mathfrak{I}} \wedge y \in C^{\mathfrak{I}}\}$	$\exists$ located_in.Country
value restriction	$\forall R.C$	$\{x \mid \forall y.\langle x,y \rangle \in R^{\mathfrak{I}} \Rightarrow y \in C^{\mathfrak{I}}\}$	$\forall$ flows_through.Country

names [Baader et al., 2004]. Defining more constructors can provide more expressive descriptions. Well-known extensions of $\mathcal{ALC}$ (abbreviated as $\mathcal{S}$) are $\mathcal{SHIQ}$, $\mathcal{SHOIQ}(D_n)$, and $\mathcal{SROIQ}$. Their syntax and semantics are given in table 2.2.

Description logic systems nowadays support different reasoning tasks. They can automatically deduce implicit information from explicitly represented knowledge structure, and they structure knowledge by using subsumption. They classify the hierarchical relationships (inheritance) among categories. They can analyze whether a specific TBox is consistent and satisfiable and whether a specific ABox satisfies the constraints given by the TBox. Such DL reasoners yield a correct answer in finite time. Question answering is also supported by many DL reasoners, i.e., particular queries about sets or truth conditions can be answered [Baader et al., 2007].

Today's standard ontology language is based on DLs and is implemented by the web ontology language OWL 2 [Motik et al., 2008]. This language is supported by different editors, such as Protégé [Rubin et al., 2007] or the NeOn (Network Ontologies) toolkit [Haase et al., 2008], as well as reasoning engines, such as Pellet [Sirin et al., 2007] or HermiT [Motik et al., 2009].[5] OWL 2 currently supports the expressiveness of the description logic $\mathcal{SROIQ}$ [Horrocks et al., 2006], which is shown in table 2.2. Ontologies can also be specified by more expressive logics, such as Common Logic [Common Logic Working Group, 2003] or CASL [Mossakowski et al., 2008]. However, these tend to

---

[5]An overview of OWL editors and reasoners is available at `http://www.mkbergman.com/862/the-sweet-compendium-of-ontology-building-tools` (visited on July 05, 2010) and `http://www.cs.man.ac.uk/~sattler/reasoners.html` (visited on July 05, 2010) respectively.

Table 2.2: **DL syntax for** $\mathcal{SHIQ}$, $\mathcal{SHOIQ}(D_n)$, **and** $\mathcal{SROIQ}$. Constructors, their syntax and semantics with examples for the description logics $\mathcal{SHIQ}$, $\mathcal{SHOIQ}(D_n)$, and $\mathcal{SROIQ}$ that extend the constructors given in table 2.1 for the description logic $\mathcal{ALC}$.

DL Constructor	Syntax	Semantics	Example
		$\mathcal{SHIQ}$	
qualified number restriction (min)	$\geq nR.C$	$\{x \mid |\{y.(\langle x,y \rangle \in R^{\mathcal{J}} \wedge y \in C^{\mathcal{J}})\}| \geq n\}$	≥ 1 flows_through. Country
qualified number restriction (max)	$\leq nR.C$	$\{x \mid |\{y.(\langle x,y \rangle \in R^{\mathcal{J}} \wedge y \in C^{\mathcal{J}})\}| \leq n\}$	≤ 1 located_in. Country
qualified number restriction (cardinality)	$= nR.C$	$\{x \mid |\{y.(\langle x,y \rangle \in R^{\mathcal{J}} \wedge y \in C^{\mathcal{J}})\}| = n\}$	$=1$ located_in. Country
inverse role	R^-	$\{\langle x,y \rangle \mid \langle y,x \rangle \in R^{\mathcal{J}}\}$	flows_through$^-$
transitive role	$^{(+)}R$	$R^{\mathcal{J}} = (R^{\mathcal{J}})^+$	$^{(+)}$located_in
		$\mathcal{SHOIQ}(D_n)$	
nominal	$\{x\}$	$\{x^{\mathcal{J}}\}$	$\{$Italy$\}$
concrete domain	$f_1,\dots,f_n.P$	$\{x \mid P(f_1^{\mathcal{J}},\dots,f_n^{\mathcal{J}})\}$	average_width $<$ 100
		$\mathcal{SROIQ}$	
disjoint role	$\mathsf{Dis}(R,S)$	$R^{\mathcal{J}} \cap S^{\mathcal{J}} = \emptyset$	Dis(located_in, has_Capital)
reflexive role	$\mathsf{Ref}(R)$	$\{\langle x,x \rangle \mid x \in \Delta^{\mathcal{J}}\} \subseteq R^{\mathcal{J}}$	Ref(part_of)
irreflexive role	$\mathsf{Irr}(R)$	$\{\langle x,x \rangle \mid x \in \Delta^{\mathcal{J}}\} \cap R^{\mathcal{J}} = \emptyset$	Irr(adjacent_to)
symmetric role	$\mathsf{Sym}(R)$	$\{\langle x,y \rangle \in R^{\mathcal{J}} \mid \langle y,x \rangle \in R^{\mathcal{J}}\}$	Sym(adjacent_to)
antisymmetric role	$\mathsf{Asy}(R)$	$\{\langle x,y \rangle \in R^{\mathcal{J}} \mid \langle y,x \rangle \notin R^{\mathcal{J}}\}$	Asy(located_in)
negated role assertion	$\neg R$	$\{\langle x,y \rangle \mid \langle x,y \rangle \in \neg R^{\mathcal{J}}\}$	$\neg$located_in (Rome, Spain)
complex role inclusion	$R \circ S \sqsubseteq R$	$\{\langle x,z \rangle \mid \langle x,y \rangle \in R \wedge \langle y,z \rangle \in S\}$	located_in $\circ$ on_continent
universal role	U		
local reflexivity	$\exists R.\mathsf{Self}$	$\{x \mid \langle x,x \rangle \in R^{\mathcal{J}}\}$	supports.Self

become undecidable. More details about OWL 2 are given in section 4.2, in which the language is used to formalize spatial ontologies.

2.3.2 Principles and Methods for Ontology Development

Logical criteria for specifying ontologies are just one aspect in ontology development, i.e., selecting the language, its expressiveness, and tool support required by the ontology. The other central aspect is deciding on the quantity of information that is to be specified by the ontology and how this has to be done. That is: *design decisions* for the ontology have to be made.

Which information is captured by a single ontology also depends on the category the ontology itself falls into. Gómez-Pérez et al. [2004], for instance, distinguish KR ontologies, *general* (also called *common*) ontologies, *top-level* (also called *upper-level*) ontologies, *domain* ontologies, *task* ontologies, *domain-task* ontologies, *method* ontologies, and *application* ontologies. The Ontology Metadata Vocabulary [Palma et al., 2009] currently distinguishes *core, upper-level, domain, task,* and *application* ontologies. KR ontologies are developed according to a specific KR paradigm, common ontologies aim at describing common-sense knowledge, upper-level ontologies provide very general types of categories and relations, domain ontologies describe only information about a specific area (e.g., baseball, wine, or music), core ontologies describe general categories of a specific domain (e.g., chemistry or biology), task ontologies specify only tasks (e.g., diagnosing, planning, or scheduling), domain-task ontologies refine task ontologies for particular domains, method ontologies formulate reasoning processes of specific tasks, and application ontologies are specialized to support specific applications they are developed for. The boundaries between these types can, however, overlap, and often the basic distinction between domain, core, and upper-level (or foundational) ontologies is sufficient for ontology development. Reusing upper-level ontologies can assist in the development of core and domain ontologies; reusing core ontologies can assist in the development of domain ontologies. The ontology type then determines the generality or specificity of categories and relations defined in the ontology. This provides a general range of the depth of information in an ontology. Concrete terms and definitions then have to be specified within this range.

The ontology development process as introduced by Uschold and Gruninger [1996] consists of the following components in sequential order: (i) identification of the purpose and scope of the ontology, (ii) formalization of the ontology, i.e., (ii.a) knowledge capture by questioning experts or analyzing the domain, (ii.b) ontology specification, and (ii.c) ontology reuse, (iii) evaluation of the ontology by analyzing requirements specification and competency questions, and (iv) documentation and guidelines about the ontology and each step in the development process. This cycle is refined in the following way [Uschold and Gruninger, 1996]: Motivation scenarios are described in the first stage, laying out possible use cases the ontology is supposed to support. Informal and subsequently formal competency questions are formulated. Such competency questions [Gómez-Pérez et al., 2004] are questions that should be answered by the ontology. In addition, knowledge of expert users can support the selection of important categories and relations to be described.

Knowing what has to be specified by an ontology leads to its formalization, which has to fulfill the given (informal) requirements. The ontology development then has to take into account expressiveness and complexity issues as well as technical considerations, implementation or tool support, and existing standards. As the formalization intends to reflect information defined by the requirements, certain design decisions have to be made by the developers. Gruber [1993] points out five design criteria: (i) clarity, i.e., ontology definitions should be complete, objective, and documented, (ii) coherence, i.e., inferences from an ontology should be consistent with the definitions, (iii) extendibility, i.e., ontology definitions should be extendable without revision, (iv) minimal encoding bias, i.e., an ontology should not depend on the language it is formalized in, and (v) minimal ontological commitment, i.e., an ontology should make as few claims as possible and provide only essential aspects about its domain [Guarino, 1998].

Several techniques and methods for the development process and lifecycle of ontologies have been proposed [Gómez-Pérez et al., 2004], which are similar to the method introduced in Uschold and Gruninger [1996] or specialized for particular tools. Among them are Methontology [Fernández-López et al., 1997], with its development stages specification, conceptualization, formalization, implementation, and maintenance; or On-To-Knowledge [Sure et al., 2003], with its development stages feasibility study, ontology kickoff, refinement, evaluation, and maintenance. Both methods allow cycles and iterations within their development process. All methods for ontology development have in common that they define different process stages, starting from a (semi-)informal description of aims and requirements of the ontology and ending with the (pre-)final version of the technical specification of the ontology. In between, different sub-stages are described together with cycles and evaluation phases. Some of the methods are constrained to be used with specific software tools (e.g., Methontology). The identification of categories may follow bottom-up, top-down, or middle-out approaches, i.e., generalizations or specializations or a mixture of both [Gómez-Pérez et al., 2004].

Ontology development methods also adopt methods from software engineering by using its development stages, iterations, and development techniques. Nicola et al. [2009], for instance, present an approach that is based on the Unified Software Development Process (UP) [Scott, 2001]. They take into account software and ontology engineering procedures by combining the four UP phases (inception, elaboration, construction, and transition) together with five workflows, namely requirements, analysis, design, implementation, and test. As part of their workflows they use storyboards and competency questions to analyze the requirements and they use UML as a modeling language before implementing the ontology in OWL. Throughout their workflows they also consider the involvement of different user groups, namely knowledge engineers and domain experts, in the development process.

In contrast to these development procedures that have a clear structure of process stages for creating new ontologies, approaches in ontology learning from data sources have also been used for generating ontologies. Thus, they can also be seen as ontology development tools. Here, categories and relations are automatically extracted from text or other digital data by linguistic and statistical analyses of the (textual) data. Because of the strong relationship to the data used, the resulting ontologies are often used for text clustering and classification [Cimiano, 2006]. Results from ontology learning can then be applied directly to new data to be clustered or classified. Refinements or revisions of automatically learned ontologies can also be part of the ontology development process itself and support its evolution [Völker et al., 2008]. Ontology learning approaches can, however, only be applied to areas for which large amounts of data exist (e.g., web documents) that can be automatically analyzed.

Recently, design patterns for ontologies have been proposed, developed in the NeOn project,[6] which are accessible online.[7] Design patterns can be reused or applied for cer-

---

[6]http://www.neon-project.org (visited on July 05, 2010)

[7]http://ontologydesignpatterns.org (visited on July 05, 2010)

tain modeling problems and can thus support design decisions during the development phase of ontologies. NeON further suggests the use of meta-descriptions, also called *meta-information*, such as an ontology's purpose, justification, competency questions, domains, scenarios, which provide general information and improve ontology usability (see also section 2.3.3). These patterns are, however, not as established as their counterparts in software engineering [Gamma et al., 1995]. The basic idea of reusing (established) ontologies on a related or general topic can help in the development of new ontologies nevertheless.

A detailed overview of many approaches for ontology building methods and procedures [Kehagias et al., 2009] as well as tutorials on how to specify ontological information is available.[8] Different tools and extensions that support ontology development and methods can be applied (cf. section 2.3.1). In summary, different approaches exist, which provide development phases for ontologies including support for distributed and collaborative ontology development and versioning. They differ slightly in the phases defined and concrete methods used, such as competency questions, storyboards, or spreadsheets. Beyond technical and methodological considerations, also content-based ontology patterns, e.g., how to model PartOf, Participation, or Plan categories and relations, are available and can be reused (see also section 2.3.3).

Although specific modeling decisions have to be made for each ontology with regard to its purpose and aims, the following strategy is applied throughout this thesis for the development of spatial ontologies: For newly defined ontologies, their requirements, purpose, aim, and application examples are defined. Related ontologies are then investigated, whether they can be used or reused directly. Specification issues, such as conventions, ontology language, application requirements, and development tools are determined. During the implementation phase the original requirements and specifications may be revised based on testing and adjustments. This cycle resembles development approaches presented by Nicola et al. [2009] and Castro et al. [2006]. However, the development procedure of the spatial ontologies presented in this thesis does not always follow this cycle. Their sometimes strong cognitive bias demands a more flexible development, which is described throughout section 4 where applicable.

---

[8]http://semanticweb.org/wiki/Ontology_Engineering#Ontology_Building_Methodologies (visited on July 05, 2010)

Development processes contribute to the match between the intended specification and the ontology result. An evaluation, however, has to analyze the appropriateness and quality of an ontology. This evaluation can be done separately from any development process or as a final phase of it. It is also applicable to ontologies that change due to revisions, new versions, or updates.

2.3.3 Ontology Evaluation

Although some methodologies for ontology development contain an evaluation phase, the research area of *ontology evaluation* has become a field of its own [Obrst et al., 2007]. Several methods for analyzing the quality of an ontology have been proposed. In this thesis, these methods are categorized according to their type of analysis. As a result, an ontology evaluation method can evaluate (i) the lexical quality of an ontology's vocabulary, (ii) the structural quality of an ontology's formalization, (iii) the semantic quality of an ontology's representation, (iv) an ontology's quality with regard to an application it is used in, (v) an ontology's accessibility and usability, and (vi) an ontology's quality with regard to philosophical considerations. These different features can be evaluated either automatically by a tool or by voting or peer-reviewing. The distinction of evaluation methods into the six quality levels is based on Kehagias et al. [2008] and Brank et al. [2005]:

Lexical Quality. Lexical evaluation methods are used to analyze whether terms of an ontology are formulated on the basis of intelligibility: term names and descriptions should not contain circular definitions, and they should be formulated in an affirmative, syntactically correct, and consistent way, in particular, by following naming conventions. Names should also comply with existing naming standardizations of the given domain. Lexical quality improves usability and readability of an ontology. Technical support for lexical quality is often provided by common ontology editors. Metrics for circularity and intelligibility of term descriptions are, for instance, presented in Köhler et al. [2006]. High lexical quality can make the use of ontologies easier, although usage is often assisted by an interface for browsing or searching ontology terms. Lexical aspects are, however, less relevant for implementation issues. The ontologies developed in this thesis ensure lexical quality primarily by providing meta-information (meta-descriptions) about the ontologies and their specification details, and they also comply with naming conventions in DLs (see section 4.2).

Structural Quality. Structural evaluation methods are used to analyze axiomatic formalizations of ontologies. Given the taxonomy of an ontology as a graph structure, for instance, its size, depth, breadth, density, balance, complexity, and connectivity, as used in graph theory [Harary, 1969], can be investigated. Well-connected and complex graph structures, for instance, imply richness of relationships, attributes, and inheritances of an ontology. Metrics for graph-based properties of ontologies are, for instance, presented by Alani and Brewster [2006] and Tartir et al. [2005]. Structural quality also analyzes expressiveness, reasoning practicability, and query simplicity, which is often determined by the language in which the ontology is formulated. Modularization as a structuring technique can also be analyzed (see section 2.4). Structural quality not only indicates the clarity and elegance of an ontology specification; it also allows a semantic interpretation of categories and relations, e.g., important concepts may have high dependencies to other parts of the ontology. It can also indicate use and reuse capabilities of ontologies, e.g., by modularization. Measuring structural quality, however, may result in different interpretations. Well-connected ontology categories, for instance, can reflect a rich or an ill-defined specification, and so has to be subject to a peer review. The ontologies presented in this thesis have been designed in a modular way (see section 4.1) in order to guarantee high flexibility for use and reuse. Their expressiveness and reasoning practicability is determined by the language in which they are formulated (see section 4.1.2). Their graph-based properties are analyzed where appropriate (see section 4.2).

Semantic Quality. Semantic evaluation methods aim at analyzing how well the ontology specifies a given domain. The ontological specification is thus supposed to be as *cognitively adequate* as possible with regard to the domain. This criterion is primarily analyzed by using peer-review results. As the ontology specification is supposed to resemble the domain it defines, its interpretation should be clear, accurate, and explicit with respect to a given context or background knowledge and avoid ambiguity. The level of granularity, i.e., detail of the specification, should be consistent throughout the ontology. The ontology should either be prescriptive or descriptive with regard to the domain and avoid a combination of both prescriptive and descriptive views. Semantic quality aspects are thus difficult to evaluate, as they are strongly influenced by aims and purposes of, as well as user satisfaction and agreement about, the ontology. However, some metrics have been proposed for interpretability, clarity, and comprehensiveness [Burton-Jones et al., 2005] and for agreement and user-satisfaction [Gangemi et al., 2005]. Both approaches are based mostly on semiotic methods, such as metrics for meaningfulness, clarity, comprehensiveness, or relevance of terms. Semantic quality of the ontologies developed in this thesis is evaluated by using peer-review methods and applying standardized domain models (see section 6.1).

Application Quality. Application evaluation methods are used to analyze the appropriateness and applicability of an ontology with regard to a certain application or data classification task. Methods used are mostly taken from text processing tools, which are based on statistical methods, such as precision, recall, and accuracy measures [Salton, 1989, p. 248]. Here, the ontology is evaluated in terms of an application or task it has to perform, e.g., classification or query answering. The results of the ontology with regard to these tasks are compared either with expected results or results of a gold standard. For data analysis, the congruency between an ontology and related corpus terms can also be analyzed. These methods are implemented and used for analysis [Brewster et al., 2004; Porzel and Malaka, 2004], however, the results vary according to a given task or an intended use, i.e., the quality of the data set affects the quality of the outcome. Hence, the data used for the ontology evaluation also needs to be statistically analyzed, e.g., whether it is expressive enough to measure the performance, or whether a predefined gold standard is appropriate. As the ontologies developed in this thesis are intended to be used within spatial systems, their applicability is evaluated by analyzing their performance in the respective applications (see chapter 6).

Accessibility Quality. Accessibility evaluation methods are used to analyze the availability and usage conditions of ontologies. Easy and effective access to ontology sources as well as a complete documentation together with annotations about the contents of the ontology support its usability. Here, meta-information [Palma et al., 2009] can also be useful for access and recognition. This also includes information about modifiability (including licensing terms), reusability (including technical interfacing), and availability (such as permanent online access) of ontologies. Metrics for annotation analysis have been presented in Gangemi et al. [2005] and metrics based on empirical analysis have been presented in Supekar et al. [2004]. However, accessibility also depends on intended purposes of ontologies, for instance, if an ontology is not supposed to be accessed directly by human users but through a specific interface system. The ontologies presented in this thesis contain meta-information for usability and they are accessible online (see section 4.2).

Philosophical Quality. As introduced in section 2.2, philosophical considerations can guide the development process of ontologies. Hence, it can be analyzed whether an ontology follows its original guidelines. For instance, the OntoClean methodology [Guarino and Welty, 2002] provides this kind of analysis based on a general categorization into the types essence/rigidity, identity, unity, and dependence. Rigid or essential properties determine categories of which every instance of the category necessarily has to be an instance of this category at any time. Identity conditions determine which instances necessarily have to be the same because they have identical properties. Unity determines categories of which every instance consists of members that are only connected to each other but not to other entities. Dependence determines whether a category defines only instances that are necessarily depending on instances of other categories. During the development phase of an ontology, each category has to be classified as one or more of these types. OntoClean's aim is to detect contradictions inherent in the specification by using this categorization and subsumption. Although it is easy to implement the OntoClean methodology [Guarino and Welty, 2002], it might be difficult to use as the exact distinction between the four categories is not always easy to apply. The ontologies developed in this thesis are generally guided by the OntoClean distinctions and foundational ontologies that are inspired by philosophical aspects (see section 4.1.1).

In summary, lexical quality of the spatial ontologies developed in this thesis is achieved by complying with naming conventions in DL (see section 4.1), and structural quality is evaluated with regard to the ontological specifications (see section 4.2). The semantic and application quality of the spatial ontologies is analyzed where applicable (see section 6.1). All ontologies are accessible online, and ontology development guidelines have been followed (see section 4.2).

2.4 Modularity

Modular design has been applied to the field of software engineering for more than 40 years. There, its aim is to reduce complexity by developing software in modular parts. Modules are thus seen as units of larger systems. Elements within one unit are highly connected with each other but they are rather weakly connected with elements from other units [Baldwin and Clark, 2006]. Hence, a complex system provides a structure that allows independence and integration of modules. Another major motivation behind modular software design is the idea of abstraction, information hiding, and interfacing:

> "A complex system can be managed by dividing it up into smaller pieces and looking at each one separately. When the complexity of one of the elements crosses a certain threshold, that complexity can be isolated by defining a separate abstraction that has a simple interface. The abstraction hides the complexity of the element; the interface indicates how the element interacts with the larger system." [Baldwin and Clark, 2006]

As a result, modules can be developed and tested independently from other modules and their interplay with the other modules can also be analyzed independently. The idea is to combine functionalities from different modules and reuse them in a complex design, allowing for flexible and easy augmentation (adding) and exclusion (removing) of modules. All these aspects improve ontological modeling in the same way.

Early work in modular ontologies has shown that modularity is seen as the crucial method for structuring large ontologies particularly to improve reuse, maintainability, and evolution [Rector, 2003]. Modularization also supports the use of ontologies in distributed systems, i.e., each system can use only those ontologies necessary for its tasks and requirements, resulting in a more efficient reasoning technique, i.e., reasoning is done only over those ontologies that are used [Stuckenschmidt and Klein, 2003].

As one of the first approaches, Rector [2003] proposed four criteria a modular DL domain ontology has to satisfy: (i) it should contain only one top node; (ii) its hierarchical definitions should be based on subsumption; (iii) it should make a distinction between primitive self-standing categories and refining categories that are properties of self-standing categories; and (iv) it should avoid cross-classifications, i.e., no category may be subsumed by more than one category. As a result, the ontology is developed in a 'normalized' form and supports that newly added categories, restrictions, or axioms will cause no inconsistencies [Rector, 2003]. The first two criteria follow automatically from the definition of ontologies and from using DL TBoxes as a formalization; the last two criteria are supported by formal ontology design guidelines (see section 2.3). As the ontologies developed in this thesis are primarily DL TBoxes (see section 4.1.2) and as they are developed with respect to formal ontology design (see section 4.1.1), they consequently satisfy these requirements. The four criteria, however, reflect merely a basic understanding of modularity, and although they are generally important more recent approaches in modular ontology design become increasingly elaborate and complex.

Nowadays modularity, in its different shapes and forms, remains one of the central research topics in ontology engineering [Kutz et al., 2010a]. Early tools for modularity mostly aimed at merging (see below) ontologies [Noy and Musen, 2002] or at supporting distributed development of ontologies [Gómez-Pérez et al., 2004]. More recent approaches investigate ontology modularity in a broader as well as more precise way: Parent and Spaccapietra [2009] define modularity as a technique for complexity management, in which each ontology module is a small ontology that specifies a sub-domain of a given domain. The module can then also define inter-module links to other sub-domain ontologies, which together yield the specification of the whole domain. Developing smaller, sub-domain modules not only reduces complexity but also fosters understandability, scalability, reuse, and interoperability. It also provides an accurate extraction and synthesis of relevant knowledge.

> "The potential specificity of modules, making them different from the whole, is that a module is aware of being a subset of a broader knowledge organization, and therefore knows it may interact with other modules to get additional knowledge in a sub-domain closely related to its own subdomain." [Parent and Spaccapietra, 2009, p. 10]

The challenging task then is to reconcile ontologies from different sources that reflect different parts of one domain, to determine and specify interrelations between such ontologies, to extract relevant parts of an ontology, and to re-combine small ontologies in order to form new and larger ones. However, the distinct modules and their interrelations can differ structurally, semantically, or functionally. Thus, mechanisms for easy and flexible reuse, generalization, structuring, maintenance, design patterns, and comprehension have to be developed and provided. Modularity is essential not only to reduce the complexity of understanding ontologies, but also in maintaining, querying and reasoning over modules [Kutz et al., 2010a].

Recent research in modular ontologies deals primarily with building large and complex ontologies by combining smaller modular ontologies based on purpose-dependent, logically versatile criteria. Such purposes include logical integration of different ontologies, association and information exchange between ontologies, the detection of overlapping parts, traversing through different ontologies, the alignment of vocabularies, or the extraction of modules. This includes the use of different logics across those parts and different approaches to types of relations between parts. Conversely, the decomposition of an ontology into sub-ontology modules is also investigated.

Stuckenschmidt et al. [2009] give a thorough overview of the breadth of modularity in ontological engineering in terms of structural representations of modularity and their various logical aspects. In this thesis, the following techniques for relating ontology modules, illustrated in figure 2.3, are distinguished [Kutz et al., 2008; Hois et al., 2010]:

Matching and Alignment. Different ontologies can be matched if the ontologies share thematically overlapping parts by defining the same entities. Their categories and relations can thus be mapped with each other. This mapping reflects identical categories and relations, i.e., it identifies the same parts from different ontologies. As a result, information or data from different ontologies can be exchanged and extended. Identified (i.e., mapped) terms across ontologies are aligned. Alignments are often established by using statistical methods and heuristics, such as similarity measures and probabilities [Euzenat and Shvaiko, 2007, p. 117 ff.].

Merging. Different ontologies can be merged with each other. It is not required, however, that their domains thematically overlap. As a result, a new ontology is

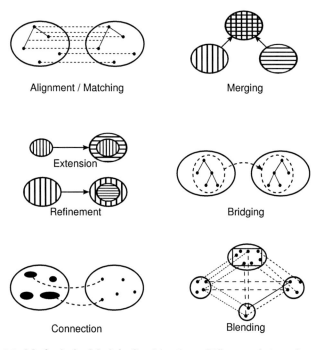

Figure 2.3: Methods for Module Combinations. Different techniques for combining ontology modules.

created that contains all information from the different ontologies. The merge, however, has to provide a solution for possible inconsistencies or mismatches between both ontologies [Hameed et al., 2004]. The axiomatization from the different ontologies are preserved in their merge, i.e., the different ontologies remain unaltered and what was defined in the original ontologies is also defined in the resulting ontology.

Extension and Refinement. If one ontology is structurally entailed within a second ontology, the second ontology refines or extends the first. Not always are the reused parts of an ontology directly or explicitly identified by the second ontology, e.g., re-namings are possible. In this case, automatic detection can yield the reused modules [Normann, 2009], and systematic structuring techniques can be used to integrate information from different ontologies and preserve inferences and implications from the reused ontologies [Kutz et al., 2008]. A specific case of refinement is a *conservative extension* of an ontology. It completely and independently specifies its own vocabulary, however, all assertions that follow from the conservatively extending ontology already follow from the extended ontology [Cuenca Grau et al., 2007].

Bridging. Bridge rules define mappings from a source ontology to a target ontology. The rules are defined separately from both ontologies, and they provide a mapping interpretation from the source to the target ontology, i.e., how categories and relations from the source ontology fit into categories and relations in the target ontology [Bouquet et al., 2004]. This formalism does not merge the contents of the ontologies, instead it provides a translation from one into the other. Although it is similar to alignments, bridge rules do not necessarily define identical categories or relations across different ontologies.

Connection. Heterogeneous connections define relationships across different ontologies while keeping the related ontologies separate [Kutz et al., 2004]. These relationships are particularly designed for connecting diverse and arbitrary ontologies of a domain or related domains, and they are not only applicable to ontologies but also to heterogeneous logics. They are thus more expressive and flexible than bridge rules by providing couplings across domains and logics and not only translations from one ontology into another.

Blending. A borderline case of relating different modules is conceptual blending [Fauconnier and Turner, 2003]. The blending of different ontologies results in a new, creatively combined ontology, i.e., new structures emerge in this new blended ontology [Pereira, 2007; Hervás et al., 2007]. The blending is based on a general base ontology that reflects information common in the different ontologies that are blended. The domains described by the different ontologies may be thematically distinct, and as a result, ontological blending can generate new ontologies and categories, which allows the most flexible technique for ontology combination compared to the previous methods [Hois et al., 2010].

In this thesis, a variety of the above methods are used to combine ontology modules for spatial information. The methods differ in logical as well as content-based ways. The choice for certain methods mostly depends on thematically relevant or application-specific requirements. In section 5.1, the combination techniques relevant for combining different modular ontologies for space are presented and applied.

2.5 Uncertainty

As pointed out briefly in section 2.1, conventional knowledge representations are specified by using a crisp logic or language. Statements about a domain are thus either true or false. In many situations, such as daily life scenarios and consequently real-world applications, the exact truth values of certain statements are not available and different notions of *uncertainties* are required. As the ontologies and their applications presented in this thesis reflect spatial real-world scenarios, they are affected by different types of uncertainties accordingly (see section 5.2). This section summarizes distinctions in uncertainty types, their general causes, and their relations to knowledge representation and ontologies.

Uncertainties are primarily a result of errors in so far as the actual information is not available for some reason [Halpern, 2003], i.e., certain facts are passively not known by an agent or system and they are not actively ignored on purpose.[9] In particular, "information which quantitatively and qualitatively is appropriate to describe,

---

[9]It can be omitted whether uncertainty is a real phenomenon that is actually present in the real world, as uncertainties can be considered to be an existing aspect for an agent or a system [Zimmermann, 2000].

prescribe or predict deterministically and numerically a system, its behavior or other characteristica" [Zimmermann, 2000] is not accessible.

Smithson [1989] distinguishes three specific types of uncertainties, namely vagueness, probability, and ambiguity. Vagueness reflects unclear boundaries of statements or unknown facts that are borderline cases. Here, it is not possible to determine precisely when a vague expression changes from true to false or vice versa. Probability indicates the likelihood of truth values of situations or expressions often using quantitative values within a range of 0 to 1. Vague statements hold to a certain degree, also measurable with values between 0 and 1, and probable statements are to be true or false with a certain likelihood. Ambiguity is caused by terms with different meanings, i.e., it depends on the interpretation of the term whether a situation or expression is true or false. In contrast to vagueness and probability, ambiguous expressions thus reflect the possibility of different term interpretations.

These three cases provide a general categorization of unknown facts into different types of uncertainties. Furthermore, Zimmermann [2000] characterizes the following reasons that can cause uncertainty in a system:

Lack of Information. Information is not available or it is too expensive to gather.

Abundance of Information. Too much information is available and it is not possible to process all of it.

Conflicting Evidence. Available information supports different hypotheses.

Ambiguity. Available information contains linguistic terms with multiple meanings.

Measurement. Available information is affected by inexact measurement of physical features.

Belief. A human being develops information by him or herself and takes it as objective information.

Hence, lack and abundance of information leads to vague or probable facts, conflicting evidence leads to probable facts, ambiguity leads to ambiguous facts, measurement leads to vague and probable facts, and belief leads to vague, probable, or ambiguous facts. Uncertain information of all kinds can be measured numerically, interval-based, linguistically, or symbolically. Numerical values can be classified according to their scale level. This is important as not all numerical values can be directly interpreted as cardinal or absolute. Permissible operations for such scale levels can be defined by having one of the scales nominal, ordinal, interval, ratio, or absolute [Zimmermann, 2000].

Numerous uncertainty theories exist that address these different types. Among them are various probability theories, evidence theory, possibility theory, fuzzy set theory, similarity theory, and non-monotonic logics. All these theories provide different methods for representing and reasoning with uncertainties, as briefly introduced in the following.

Probability Theories. Probabilities indicate how likely it is that an event or a statement is true or false [Bhatnagar and Kanal, 1986]. Probability values assign a likelihood degree between 0 and 1 to an expression, i.e., if a statement is likely to be true to 40 % then its probability is 0.4. Probabilities also enable the addition of uncertainties, i.e., given the probability of different statements the probability of their combinations can be calculated. A well-known example of a probability theory is Bayes theory, in which Bayes' Rule defines how conditional probabilities can be calculated and how probability distributions over statements can be defined [cf. Hartigan, 1983]. Probabilities are based on statistical data of the frequency of occurrences of events, i.e., probability values can be measured and gained from a data set. Thus, probabilities are classified as *objective* uncertainties [Zimmermann, 2000].

Evidence Theory. Evidence theory is a refinement of probability theory. It relates statements to a degree of belief, i.e., how likely they are true or false [Shafer, 1976]. This belief, however, is defined with regard to a human or an agent intended to support the open world assumption, i.e., the number of statements is infinite. Beliefs can be combined and a belief distribution over statements can be calculated. Agents or systems can gain their beliefs on the basis of epistemic data, i.e., subjective judgments from an expert determine actual belief values. Hence, evidences are classified as *subjective* uncertainties [Zimmermann, 2000].

Possibility Theory. Possibility theory provides two measures for the likelihood of an event, namely its possibility and its necessity. In both cases, a number between

0 and 1 is assigned to sets of statements from a finite number of events. Whereas in probability theory, probabilities can be calculated based on their conditional dependencies, in possibility theory, combinations of possibilities and necessities are maximized and the different sets of statements can be ordered, resulting in a weighted logic of uncertainty [Dubois and Prade, 1988]. The combinations then yield a possibility distribution.

Fuzzy Set Theory. Different from the previous uncertainty theories, fuzzy set theory defines the degree to which a statement holds [Zadeh and Kacprzyk, 1992]. Hence, a statement is not likely to be true or false, but it is true to a certain degree. This kind of uncertainty measurement is often used for dealing with vague natural language terms. Terms such as warm and cold or forest and mountain can be said to hold to a certain degree in fuzzy set theory.

Similarity Theory. Similarity theories can be used in order to specify concept resemblance [Tversky, 1977]. The similarity between concepts (or statements) can be described by means of their features. The more similar features different concepts have, the more similar are the concepts to each other. However, geometrically-based similarities are another way to define similarities between concepts, e.g., similarities between conceptual spaces [Gärdenfors, 2000]. Furthermore, similarity can be measured by using proximity values between categories [Sheremet et al., 2007]. While 0 indicates closest similarity, no similarity is indicated by infinity.

Non-monotonic logics. Logics for representing uncertainty extend classical logics by either introducing more truth values or more operators or inference mechanisms [Brachman and Levesque, 2004]. Among them are three-valued logics, autoepistemic logics, or default logic. The latter, for instance, is used to describe general statements [Reiter, 1980], e.g., normal, prototypical, or statistical statements. It basically determines what it is consistent to assume for a given statement. Although non-monotonic logics are not classified as uncertainty theories, they provide a mechanism to represent and reason with uncertain information.

Although this list gives a general overview of the variety of uncertainty theories, it is not exhaustive. Concrete applications that have to deal with uncertainties may prefer one method over another. This decision is mostly influenced by the requirements of each application individually. [Brachman and Levesque, 2004]

In summary, probability theories provide a way to reason with objective probabilities based on statistics, whereas belief theories provide a way to reason with subjective probabilities based on expert knowledge. In order to determine the likelihood of an event, the former theories are more applicable when large data sets are available and the latter theories are more applicable when no data sets but expert knowledge is available. Possibility theories aim at generalizing over the likelihood of sets of statements based on the general possibility of an event to occur, i.e., no statistical or expert knowledge is necessarily required. Fuzzy set theory provides a method to describe degrees of applicability of statements, particularly applicable when dealing with vagueness caused by linguistic data. Similarity measures are often applied when comparing categories or concepts or also linguistic meanings.

Uncertain reasoning based on one of the uncertainty theories is specifically applicable for common-sense reasoning, as crisp logics become difficult and inefficient due to the unavailability or lack of accessibility of data. When it is difficult to gauge something precisely or categorize something perfectly, different types of uncertainties as provided by the uncertainty theories have to be included in the knowledge representation [Brachman and Levesque, 2004]. In the context of spatial information systems that use ontologies to represent space, different types of uncertainties and appropriate methods for representing them are required. These are discussed and presented in section 5.2.

2.6 Representations of Space

One of the oldest representations of spatial information is most likely the system of Euclidean geometry, which is defined by geometrical axioms with points and lines as primitive entities [Hilbert, 1899; Tarski, 1959]. It has been used and applied within computational systems dealing with space. Together with numerical information available from a system's input data and the use of three or other dimensional coordinates, a system can build an internal structure of its spatial environment. However, this kind of representation is inappropriate to capture the way humans deal with space and thus the way cognitive agents are supposed to deal with space. In particular, cognitive and ontological considerations show that space is represented in many different ways [Bateman et al., 2009]. As a consequence, an ontological formalization of spatial information also need to take these different aspects into account if it intends to model space in a cognitively adequate way. An overview of approaches in different research areas that analyze spatial representations beyond geometry, namely spatial cognition, spatial logics, spatial ontologies, and spatial applications, is given in this section.

2.6.1 Spatial Cognition

The field of spatial cognition investigates the ways humans perceive, observe, understand, communicate about, reason with, interact with, and navigate through space [Ó Nualláin, 2000]. It builds and evaluates psychological or mental models (see below) of human common-sense knowledge and models for processing space and spatial information. Often models are implemented in computer or robotic systems and evaluated with regard to their performance and adequacy. These approaches are usually based on or motivated by experimental studies or empirical analyses. Ontological representations of space, in particular when applied to spatial assistance systems, i.e., systems that assist humans in spatial tasks, can thus benefit from taking into account how humans categorize, perceive, or talk about space (cf. chapters 3 and 6).

One of the first works, and also one of the most detailed, using psychological studies to investigate space is that of Piaget and Inhelder [1997]. They analyzed when and how children acquire and develop a spatial understanding, and they studied the spatial aspects children are able to distinguish. Their results indicate that human representation of, and reasoning with, space go far beyond mathematical and geometrical constructs or co-ordinate systems by using both pure perception (perceptual space) and representation (conceptual space and mental imagery). In children's early development of spatial understanding, only primitive perception-based spatial relationships are distinguished, such as proximity, separation, order, enclosure, and continuity, followed by haptic recognition of familiar and topologically different shapes. One of the major outcomes of the experimental studies is that space cannot be reduced to geometrical properties but is perceived by different senses and experienced through performing different actions abstracted and linked to object categories [Piaget and Inhelder, 1997, p. 449].

Johnson-Laird [1983] has shown that human reasoning is based on representations in analogy to spatial representations. These representations are called *mental models*. They can reflect "objects, states of affairs, sequences of events, the way the world is, and the social and psychological actions of daily life" [Johnson-Laird, 1983, p. 397]. In particular, spatial problems are solved by using mental diagrams that depict a spatial situation and that can be (mentally) processed. The mental model is supposed to be structurally identical with the spatial situation, i.e., the relations that hold in the situation can be inferred from the model.[10]

Johnson [1987] has shown that humans use imaginative but recurring patterns that enable them to structure their observations and experience while moving through and interacting with their spatial environment. These patterns are also called *image schemata*. They are supposed to be well-defined and to provide enough information to support human understanding and reasoning. They also provide a more abstract representation of an environment than mental images, however, they are less abstract than logical or formal spatial structures (see section 2.6.2). They can rather be seen as a (not only spatial) generic structure that can be applied to different environments. Examples of schemata are CONTAINER, SURFACE, LINK, PATH, NEAR-FAR, PART-WHOLE, and CENTER-PERIPHERY. Furthermore, it has been argued that metaphors, which map patterns from one domain into another, use such image schemata [Lakoff, 1990] and it has been shown how this mechanism can also be applied to user interfaces [Kuhn and Frank, 1991].[11]

Olson and Bialystok [1983] presented a thorough analysis of which types of spatial representations humans can learn, develop, and use. They particularly analyze humans' perceptual ability and how humans recognize and classify objects. The mental

---

[10]Note that there has been a long debate whether humans store their memories as images or as propositions or a composition of both [Kosslyn and Pomerantz, 1977]. As this thesis aims at a formalization of spatial information reflecting diverse types of it, it does not aim at specifying mental models of space. However, the relationship between ontological categories and their groundings to real objects is often required for spatial systems and is discussed in chapter 4.

[11]Another well-known (spatial) pattern-based mechanism is Gestalt theory [Wertheimer, 1924]. Here, complex forms can be perceived by the collection of separate visual elements, though each element alone does not indicate the complex form. Key Gestalt properties are emergence, reification, multistability, and invariance, which are supposed to be fundamental to the nature of human perception [Lehar, 2004]. As these patterns are primarily applicable to visual perception, they are less relevant for the purpose of this thesis.

representation of objects is supposed to be reflected by a propositional representation that is used for spatial cognitive tasks. Humans mentally represent objects by means of their object level and their form level. The object level provides perceptual information of object features, i.e., those features that are perceived by humans. The form level describes structural information of object features, i.e., the spatial characteristics and arrangement of objects and their parts. The combination of both levels allows the recognition of objects and the assignment of meaning to objects. Both levels are, however, tightly intertwined, and one cannot be understood without the other [Olson and Bialystok, 1983, p. 35].

The analysis of locative prepositions for spatial relationships appearing in natural language can help determine the types of spatial relations humans are able to express and conceptualize [Talmy, 1983]. Spatial linguistic features can be ordered in the way humans acquire them: (i) spatial prepositions, i.e., spatial relationships between objects, (ii) invariant properties of objects, such as the front side, (iii) complexity within usage of different types of spatial relationships, such as topological and orientation-based relations, (iv) multiple dimensions of relations, e.g., vertical and horizontal axes information, (v) different perspectives with regard to a reference object, the relatum, in the spatial relationship, and (vi) complexity within usage of the different previous features [Olson and Bialystok, 1983, p. 64 ff.]. Herskovits [1986] provides an approach to formalize English locative expressions resulting in a lexical semantics in order to determine the meaning of spatial relations, their categorization, and interpretation. The approach distinguishes two levels of abstraction, namely the ideal meaning and the use type. Although the ideal meaning is not sufficient to determine the truth values of spatial locative expressions, it provides an abstract organization of the overall set of possible expressions. The use type defines meaning derived from the ideal allowing the determination of truth conditions [Herskovits, 1986, p. 18]. The way such results can support the formulation of an ontology that contains this kind of information is shown in section 3.2.4.

Finally, theoretical and empirical findings primarily in inter-cultural studies suggest that spatial cognition is pervasive across cultures and behaviors. "Many important aspects of spatial cognitive structures and processes are universally shared by humans everywhere" [Montello, 1995]. Furthermore, Freksa [1991] points out the generality of spatial cognition: Not only do humans perceive space multimodally, namely by hearing, sight, touch, smell, and temperature sensation, but humans also transfer spatial capabilities to other cognitive skills, e.g., metaphoric descriptions. The multimodality in perception induces different types of aspects of space that humans are able to distinguish (cf. section 3.2.4). While spatial knowledge for humans is qualitative in nature, the world around them is quantitative in nature. Hence, a representational structure for space, particularly for use in a spatial assistance system, has to take both aspects into account (cf. section 3.2.1).

Humans ideally "represent only knowledge which (... they) will need for solving the tasks which (... they) have to solve and abstract from all other knowledge" [Freksa, 1991, p. 364]. This supports the position that spatial information is split up into smaller components while ignoring others. Using only necessary aspects for spatial reasoning thus requires the spatial representation to enable a selection of these necessary components, e.g., by providing modular structures that can be selected on the basis of their appropriateness (cf. section 3.1). The spatial representation, however, has to deal with (i) limited perceptual knowledge in terms of resolution, completeness, or certainty, (ii) contextual information that is not entirely accessible or knowable, (iii) a finite amount of perceptive information, and (iv) neighborhood relationships among objects that can change with the context but that can provide useful information for spatial reasoning. Hence, the different aspects of spatial information are affected by uncertainties, and humans have to cope with this problem to be able to use or reason with spatial information (see also section 5.2).

The different aspects of space as analyzed in the area of spatial cognition underly the ontological specification of space that is described in chapter 3. In particular, the different perspectives and components lead to a modular specification that is described in chapter 4.

2.6.2 Spatial Logics & Spatial Calculi

The first in-depth study of space from a mathematical perspective was the Euclidean geometry [Hilbert, 1899; Tarski, 1959]. Here, the system of geometry was defined by axioms based on points as primitive entities. It furthermore provided the ground for follow-up investigations into *spatial logics*. Spatial logics are formal languages used for describing geometrical (the term being used here in a broad sense) entities and their relations (sets of relations are also called configurations) [Aiello et al., 2007]. They are interpreted over a class of structures for geometrical entities and relations, which can be any kind of geometrical spaces, such as topological spaces, affine spaces, metric spaces, or Euclidean 3D-space.

Spatial logics are specified by a spatial (logical) language. Once this language is defined, its valid formulas, its expressive power, its computational complexity, and its alternative interpretations can be analyzed [Aiello et al., 2007, p. 5]. Relevant for this analysis are the logic's true sentences over all interpretations, its expressivity that determines the invariance relations across models and thus the level of semantic resolution, satisfiability and model-checking for complexity-theoretic analyses of the language, and the number of possible models that influence the range of applicability. Spatial logics can specify the spatial (geometry-based) entities, their relations, and their axioms in a broad way. A more particular subclass of them are logics of *qualitative spatial reasoning*. The motivation of these logics builds on the idea that reasoning over qualitative descriptions more closely resembles human reasoning and so may lead to more efficient and effective reasoning strategies than reasoning over numerical descriptions (cf. section 2.6.1). Hence, qualitative spatial representations can serve as a method for defining particular qualitative spatial terms and their relations. Within the field of *qualitative spatial representation and reasoning* (QSR) a variety of such representations have been developed so far. These are often called *spatial calculi* as they specify a *composition table*, i.e., results of combinations of spatial relations are available by syntactical inferencing.

Spatial calculi commonly focus on one particular qualitative spatial aspect, such as distance, shape, orientation, or topology, and on primitive types of objects in the Euclidean plane, such as points or regions. The selection of these aspects is based on results from empirical studies that show the way humans acquire knowledge about topology, orientation, distance, etc. [Renz and Nebel, 2007], as discussed in the previous section. Spatial calculi define possible spatial relations among entities and additional axioms together with composition tables. In many cases, *neighborhood graphs* [Freksa, 1991] can be defined on the basis of the spatial relations. In these graphs, the spatial relations of the calculus are the nodes. If a spatial relation between spatial entities can change fluently into another spatial relation without going through any other relation, their corresponding nodes are connected by links. Neighborhood graphs can be interpreted as fluent transitions between relations and can even emulate spatial motion [Freksa, 1991]. Most often, spatial calculi are also intended to serve as a representation that takes into account the diversity of space in a cognitively acceptable way [Cohn and Hazarika, 2001]. Some spatial calculi even reuse linguistic terms in order to express their spatial relationships (e.g., Krieg-Brückner and Shi [2006]); their interpretations are, however, determined by the calculus' axioms and not by linguistic constraints (this relationship is discussed in section 6.4).

Besides (or in fact, because of) their presumed cognitive adequacy, spatial calculi are supposed to provide efficient reasoning strategies, and tools are available that implement and approximate this — such as SparQ [Wallgrün et al., 2007], GQR [Gantner et al., 2008], or QAT [Condotta et al., 2006]. Here, reasoning is based on compositions and constraint satisfaction. The relations of a spatial calculus can be used to formulate spatial constraints between objects from the domain of the calculus. The resulting specification can then be formulated as a spatial *constraint satisfaction problem* (CSP), which can be solved by using applicable reasoning techniques, for instance, by applying composition and intersection operations on the incorporated relations. Qualitative spatial reasoning is used in this thesis for spatial application tasks as described in section 6.2.

Often used spatial calculi are topological theories, which are also called *Region Connection Calculi* [Randell et al., 1992]. They have their roots in the topological theory of connections between objects introduced by Clarke [1981]. The basic part of a region connection calculus (RCC) assumes a dyadic relation of *connection*, namely $C(a, b)$, denoting that region a is connected to region b, thus two well-behaved regions are related with each other with regard to their topological connection relationship. The regions are commonly regarded as regular-closed subsets of a topological space, i.e., the boundary together with the region it contains form the region. Hence, when connected the topological closures of a and b share at least one point [Renz, 2002, p. 48].

Randell et al. [1992] define the spatial calculus RCC-8 as one example of region connection calculi. This calculus defines 8 basic spatial relations (also called base relations or primitive relations) that can hold between two regions. In figure 2.4, the 8 basic spatial relations defined by RCC-8 are illustrated: disconnected $dc(a, b)$, externally connected $ec(a, b)$, partial overlap $po(a, b)$, equal $eq(a, b)$, tangential proper-part $tpp(a, b)$, non-tangential proper-part $ntpp(a, b)$, tangential proper-part inverse $tpp^{-1}(a, b)$, and

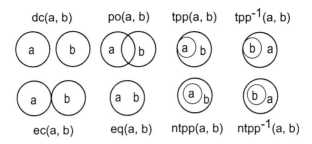

Figure 2.4: RCC-8. The 8 basic spatial relations of the spatial calculus RCC-8.

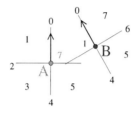

Figure 2.5: $\mathcal{OPRA}_2$. The $\mathcal{OPRA}_2$ relation $\vec{A}\ _2\angle_7^1\ \vec{B}$.

non-tangential proper-part inverse $\mathsf{ntpp}^{-1}(\mathsf{a},\mathsf{b})$. These relations build a jointly exhaustive and pairwise disjoint set of basic relations[12], which is referred to as the RCC-8 fragment of the region connection calculi. Topological spatial relations are often relevant in spatial assistance systems, thus RCC-8 has been used for the ontological specifications of space in this thesis, as described in section 4.2.1 and section 6.2.

Not only topological but also orientation-based relationships have been specified and analyzed. One of them is the orientation calculus *Oriented Point Relation Algebra with granularity m* ($\mathcal{OPRA}_m$) [Moratz, 2006], which provides a formalism for relative orientation relations between oriented points, i.e., points in the plane with an additional direction parameter. An oriented point $\vec{O}$ is defined on the basis of its Cartesian coordinates $x_O, y_O \in \mathbb{R}$ and an additional direction $\phi_{\vec{O}} \in [0, 2\pi)$ with regard to an absolute frame of reference. The parameter m in $\mathcal{OPRA}_m$ determines the angular resolution, i.e., m defines the number of basic relations.

---

[12]In the case of binary relations, 'jointly exhaustive and pairwise disjoint' limit pairs of entities to have exactly one basic relation. For n-ary relations in spatial calculi this requirement has to be satisfied for any n-tuple.

For instance, $\mathcal{OPRA}_2$ defines oriented points that can distinguish 8 different orientations, i.e., for each pair of oriented points two lines are used to partition the plane into 4 planar and 4 linear regions. An example of this partition is illustrated in figure 2.5. The orientation regions are enumerated starting from 0 to 7. The orientation region 0 always defines the (intrinsic) orientation of the point. An $\mathcal{OPRA}_2$ basic relation is a pair (i, j), in which i is the number of the orientation region from the viewpoint of A that contains B and j is the number of the orientation region B that contains A. Such relations are written as $\vec{A} \, {}_2\angle_i^j \, \vec{B}$. $\mathcal{OPRA}_2$ furthermore defines additional basic relations by which the two oriented points are located at the same position but may have different orientations ($m\angle i$). The ontological specifications presented in this thesis are applied to vision-based object recognition, in which orientation-based relations are relevant for object classifications, as described in section 6.3.

From an ontological point of view, spatial calculi provide an abstraction of space, mostly formal or mathematical. Although elements of the spatial domain are simplified as being regions or points, their relationships are defined on a fine-grained level. Further relationships can be inferred from a qualitative spatial model, i.e., from a given set of objects and spatial relations among them. As they are supposed to model human cognitive abilities, their specification can be compared with other abilities, such as spatial natural language (see section 6.4). Furthermore, their high expressivity and reasoning capability allow their implementation in spatial systems, for instance, in assisted architectural design (see section 6.2). Thus, spatial calculi are relevant for specifying modular ontologies for spatial information, as discussed in section 4.2.1.

2.6.3   Ontological Analysis of Space

In this section, several ontologies are presented that address spatial information or that model space as a category together with concrete spatial objects, spatial relations, spatial processes, or other spatial information. Some of these ontologies are reused in applications or spatial specifications or their modeling strategies or interpretations of space are applied to build ontology modules in this thesis (see section 4.2). Most of the following ontologies, which are selected by their advanced stage of development for space, mostly address spatial information from an abstract, high-level perspective, i.e., they are specified by upper-level ontologies. Although these are too general to specify ontologies for spatial information in detail as presented in chapter 4, they can provide guidelines or general modeling decisions. For instance, a basic (philosophical) distinction can be drawn as to whether an ontology follows a Newtonian (also called Galilean or absolutist) view of space, i.e., space exists independently of any objects that happen to have locations within space, or alternatively a Leibnizian (also called relational) view of space, i.e., space is mainly a matter of inter-relationships between objects [Borgo et al., 1996]. A choice for one of these views affects the way space is modeled by an ontology, i.e., either directly as a category or indirectly embedded within relations.

In SUMO, the Suggested Upper Merged Ontology [Niles and Pease, 2001], an upper level ontology that is structured into sub-ontologies, spatial information is modeled by categories that are related to physical objects, mereotopological information (i.e., parthood information), spatial relations, and attributes and functions related to the spatial domain. Spatial entities (Object) are subsumed by Physical in the SUMO hierarchy and further subdivided into Agent, Collection, Region, and SelfConnectedObject. The first two categories are defined to have a space-time position, their actual spatial information is reflected by the categories Region and SelfConnectedObject. The category Region defines a topographic location in space and its instances "encompass surfaces of Objects, imaginary places, and GeographicAreas" [Niles and Pease, 2001]. In particular, other entities (such as agents and collections) are located in a certain region, i.e., empty regions do not exist. SelfConnectedObject is identified by having only constituent parts that are connected mereologically with each other, i.e., instances of this category do not "consist of two or more disconnected parts" [Niles and Pease, 2001]. Spatial properties of entities can be specified with the category ShapeAttribute, which is subdivided into categories that define a particular geometrical form of an entity (e.g., a ClosedTwoDimensionalFigure), and with the category PositionalAttribute, which is subdivided into several particular spatial position nominals (e.g., North). Spatial relationships between objects can be specified by using mereotopological relations, similar to the spatial relations in RCC-8.

In OpenCyc,[13] the freely accessible version of Cyc [Lenat and Guha, 1990], space is modeled as a microtheory that specifies different types of spatial objects, properties, and relations. Among them are surfaces, locations, and mereotopological relations.

---

[13]http://www.opencyc.org, visited on July 05, 2010

OpenCyc aims at distinguishing the variety of relationships that are linguistically expressed by the preposition 'in', for instance, if the inner object sticks out of the outer object or not, if the inner object is between outer objects, if the inner object falls out of the outer object when it is turned around, or if the inner object is attached to the outer object (cf. also section 6.4). The ontology therefore distinguishes different types of parthood relations. These, however, also contain types that are not explicitly spatial in nature, for example, a piece of information SubInformation that is part of certain information ("Jim and Mary watch TV" has "Mary watches TV" as a part of its information). Further examples of parthood relations are Ingredient, AnatomicalPart, Constituent, and SubEvent. OpenCyc also differentiates between 63 types of shapes, e.g., Arc, RoundShape, and Amorphous. They define properties of the outer surfaces that objects can have.

In the foundational ontology DOLCE, the Descriptive Ontology for Linguistic and Cognitive Engineering [Masolo et al., 2003], space is described only at the most abstract level. Objects can have a spatial quality that can ascribe spatial attributes to the objects. In particular, this spatial attribute can, for instance, be specified by using quality spaces [Gärdenfors, 2000], i.e., a set of abstract (qualitative) spatial regions that define particular spatial aspects and that can be axiomatized internally. In detail, an object can be related to a spatial quality that is related to a particular instance of a quality space. Other spatial formalisms, e.g., quantitative or geometric spatial aspects, can also be used. However, possible quality spaces or spatial formalisms have to be defined in order to extend the DOLCE ontology, as DOLCE does not provide this information itself. Nevertheless, the general modeling pattern of entities having a spatial quality that can define a particular spatial position or region is generally relevant in formalizing space and spatial information (see section 6.3).

BFO, the Basic Formal Ontology [Grenon and Smith, 2004], is an upper level ontology that aims at reflecting information from a realist perspective, i.e., ontological entities refer directly to entities in reality. In particular, BFO specifies spatial entities by a snap and a span ontology. The snap ontology contains information about temporal aspects, i.e., it provides a snapshot of a certain domain; the span ontology is intrinsically temporal, i.e., it involves 4D views on the domain and all entities happen in space-time. In the BFO-SNAP ontology, space is defined as a category of its own, namely by the category SpatialRegion. This category acts as a location for other SNAP entities, in particular for the category SubstantialEntity. SubstantialEntity defines ordinary physical objects and parts thereof. Furthermore, the location at which these physical objects or their parts are located are defined by the category Site. Site differs from SpatialRegion as it is not based on mereotopological relationships, but it reflects more object-like information (e.g., a room in a house or a landing strip) and is related to its location by a SpatialRegion. A particular subtype of Site is the category Niche that indicates a spatial location primarily defined by functional aspects [Smith and Varzi, 1999]. In the BFO-SPAN ontology, the category SpacetimeRegion provides a location for all entities in space-time, and a space-time Site is also defined. The specification thus resembles the spatial aspects in BFO-SNAP. Furthermore, BFO defines the dimension of 'perspectivalization', which is a notion of granularity that can be treated as a specification of relative size or resolution. This reflects the (spatial) plurality in distinct ontologies, which allows zooming into and out of different scales.

As space is a fundamental category (such as time) it is often only abstractly formalized in upper-level or foundational ontologies (e.g., in DOLCE). In many cases, space is mostly addressed by using mereology and topology (e.g., in SUMO). An ontology can specify space as a category of its own (e.g., in BFO) or as a property of physical objects (e.g., in DOLCE). The upper-level ontology approaches for modeling space as described above can thus give a general orientation about space and spatial aspects and how to model them ontologically. Because of their abstract description, more precise or extensive distinctions about spatial information, for instance, those made by the approaches introduced in the two previous sections, are naturally not addressed by these upper-level ontologies. Still upper-level ontologies can guide domain-specific ontologies in their specification of spatial aspects or provide a general influence on the ontological specification [Bateman et al., 2009]. Thus, some of these ontologies or their fundamental spatial distinctions are (re-)used by the spatial ontologies presented in this thesis (see chapters 3 and 4).

2.6.4 Spatial Applications

There are several types of application that rely on spatial information. Among them are systems for geographical information, visual recognition, human-computer interaction, robotics, and design. These have different spatial tasks to perform, such as spatial problem solving, spatial inference, spatial assistance, spatial visualization, consistency checking of spatial information and maintenance, navigation in space and wayfinding, motion planning, and robot task planning. Such tasks require a highly explicit and spatially appropriate specification. The majority of approaches that apply spatial ontologies for this specification lies in the area of geographical systems, but also vision and robotics systems are increasing their use of ontologies. In section 6, applications using the spatial ontologies developed in this thesis are presented and discussed, and one main motivation for the overall work in this thesis is to provide spatial ontologies that are applicable to spatial assistance systems.

The following section outlines example applications from the fields of geography, cartography, vision, and robotics that have developed and applied space-related ontologies. It shows a sample of spatial ontologies that have been successfully used for specific application tasks, motivating the objective of this thesis. The following example applications, however, have specialized their ontologies for application tasks often without following ontological design guidelines, which makes their ontologies difficult to reuse and extend. As the spatial ontologies presented in this thesis comply with general spatial distinctions and design guidelines (cf. sections 3.2 and 4.1), they provide an application-independent and reusable spatial specification (see chapter 6).

In the area of *Geographical Information Systems* (GIS), ontologies are a suitable method to address the broad variety of spatial geographical entities and their properties [Frank, 2001]. Early GIS mostly used basic geometrical entities, such as points, lines, and areas, however, the increasing demand for specifying common-sense entities and more complex geographical elements led to the development of GIS-related ontologies. One such approach is presented by Kovacs et al. [2007]. They propose a method for specifying ontologies that formalize geographical entities. All these entities are physical topographic objects. In particular, each of them has a 'footprint', which is a projection onto a two-dimensional space. The projections can be related to polygons from a geographical data set, e.g., polygons from a map representation of the Earth's surface. The physical entities specify, for instance, land cover types, such as forests, rivers, fields, or meadows.

Maps are an often used representation format for spatial (and non-spatial) information not only in GIS but in general [Montello, 1998]. They provide a simplified visual image or network showing multiple types of information within a single structure. Maps containing primarily spatial information are cartographic maps or road maps, which may contain arbitrary types of data, such as places, elevations, street names, official buildings, or iconic entries for tourist-related information. They can show routes between two distant points and comparisons between different routes. Maps can even make invisible entities or boundaries visible, such as borders or traffic restrictions. Kuipers [1978] was one of the first to introduce maps as a spatial knowledge representation schema. His 'TOUR' model is separated into five layers, which can also be seen as an ontological analysis of map aspects. The five categories involve "route descriptions, topological street networks, coordinate frames for relative positions, dividing boundaries and grids, and structures of containing regions" [Kuipers, 1978]. These are inspired by cognitive aspects, namely human perception, human observation, and common-sense knowledge.

Ontologies can also be used to support active vision or complex scene recognition tasks. One example of such an ontological specification is introduced by Maillot and Thonnat [2008]. Their ontologies have been developed for an object recognition system to provide a classification schema for pollen grains, i.e., their ontologies primarily provide a vocabulary for the domain. Together with a learning mechanism, the ontologies are used to classify images of pollen grains. The classification is based on a training set of sample images. Based on this training set, classification-relevant dependencies between pollen grain classes and their features are extracted. The ontological specifications are based on images taken from pollen grains and their classifications from expert users (biologists).

There are four ontologies for different types of information: texture, color, space, and image acquisition context. The texture ontology distinguishes three main aspects according to textural perception, which is inspired by results from empirical studies. The three categories are repartition, contrast, and pattern, which are all further refined and quantified. The color ontology distinguishes several color variations along the dimensions hue, brightness, and saturation. These are taken from a standardized color dictionary. The spatial ontology describes geometric concepts for shape of objects, their position, orientation, and size, which can be quantified. Also, relations based on RCC-8 are defined. The image acquisition context ontology consists of meta-information about the images. It describes the date, the point of view, the location, and the sensor, and it is intended to be extended with respect to particular application purposes.

In summary, the approach combines statistical learning techniques with background knowledge provided by ontological representations. The image recognition system therefore benefits from the combination of two complementary aspects, namely ontological and statistical information. This synergy between statistical data and ontological specifications is particularly relevant for grounding ontologies (see section 4.1). The spatial ontologies presented in this thesis have also been used in the area of visual recognition (see section 6.3), showing that they can contribute to scene classifications in combination with statistical data.

In the field of robotics, robotic agents can benefit from using an ontological representation of their environment [Williams, 2008]. Explanation and prediction but also planning and collaborating with other agents is directly supported by such an abstract representation, since ontologies aim at providing common-sense and general representations of a domain. As a result, a robot is not only aware of its sensorimotor input data but it can also relate this data to an environmental or domain-specific representation that adds semantics and grounds the data. Johnston et al. [2008] present an approach of a robotic agent that uses an ontological representation in which objects in a soccer environment (RoboCup[14]) are specified. The ontology defines entities, such as balls, goalkeepers, beacons, and their properties, such as shape, size, or color. The ontological formalization not only supports the robotic vision system in object categorization tasks but also provides a vocabulary for communication between agents.

The ontologies developed in this thesis aim at a general formalization and classification of spatial information from different perspectives. However, the resulting ontological framework can be applied to a variety of different spatial applications. A selection of three domains and different tasks is presented and discussed in chapter 6 and the ontological representation is evaluated with regard to these applications.

2.7 Chapter Summary

This chapter has surveyed the state of the art in knowledge representation, ontologies, and spatial information systems. Related work in these areas relevant for this thesis was presented, and the topics to which this thesis contributes were pointed out. Technical details have been presented, namely ontology languages, modularity, and uncertainty.

---

[14]http://www.robocup.org (visited on July 05, 2010)

These are relevant for the concrete implementation of the spatial ontologies in chapters 4 and 5. Prototypical applications, use cases, and evaluations are presented in chapter 6. Before this, however, the domain space is analyzed in more detail. Hence, the next chapter presents an ontological analysis of space and spatial information, and the resulting formalization of space from different perspectives.

3

Types of Space and Spatial Information

In this chapter, we present how types of spatial information can be distinguished conceptually and ontologically. These types reflect different perspectives on space, i.e., how space can be described and categorized. We therefore introduce the term *spatial perspective* to classify the different spatial types and illustrate their differences with spatial examples. We argue that four main spatial perspectives can be distinguished, namely quantitative/qualitative, abstract, domain-specific, and multimodal perspectives on spatial information. Furthermore, a spatial perspective can be specified by an ontology module that complies with the respective spatial perspective. In this regard, we also discuss how ontological modularity directly supports the representation of multiple perspectives on a domain.

We also analyze ontological requirements of modules complying with spatial perspectives. Based on the classification of the spatial perspectives, ontology modules complying with different spatial perspectives have conceptually and ontologically distinct characteristics. Hence, the spatial perspectives structure spatial information and provide a separation of spatial ontology modules. The modules can be developed and distinguished accordingly by the spatial perspectives. Thus, the perspectives not only differentiate the types of spatial information thematically but also define requirements for respective ontology modules.

3.1 Categorizing Space

The overview in section 2.6 has shown that a large variety of different representations for spatial information exist. They range from abstract or mathematical formalizations, such as spatial calculi, to cognitive-biased models, such as image schemata. Hence, the contents and meanings of these representations can highly differ. For example, spatial representations can describe space with different detail or granularity (fine-grained or coarse), for different purposes and domains, on an abstract (thematically broad) or a specific (thematically detailed) level, or with inspirations from psychological or cognitive (spatial) models. This diversity of possible description of spatial information can accordingly result in a variety of different ontological representations for specific spatial types. Each individual ontology describes spatial information with respect to specific spatial contents and meanings, i.e., it describes the spatial domain from its own specific perspective. In this thesis, this is called the *spatial perspective* that an ontology module can comply with.

For example, an ontology module can specify the category Door from different spatial perspectives. A door can be modeled as a physical object with certain properties, e.g., the door can be open or closed, locked or unlocked, wooden or glass. A door, however, can also be modeled as an entrance to another area, such as a building, roof, garden, or car. Doors can also be specified by means of their door type, such as revolving doors, sliding doors, or wing doors. Furthermore, doors can be described as an access restriction unit that can automatically control or monitor user access. For navigation, a door can also be seen as a connecting element between two rooms. For emergency reasons, a door can be classified as a fireproof door. As a spatial artifact, a door can also be used as an information board, i.e., it can carry information or door signs. For user interaction, a door can also provide possible action types, such as opening, reading, knocking, or ventilating.

Such thematically different scenarios are illustrated in figure 3.1. An individual representation for the category Door describes a door from a specific perspective and selects necessary characteristics that are required to model the respective door type. As the different representations still model the same door in reality, the perspectives are not mutually exclusive. An individual representation, which is specified by an ontology module, can comply with more than one perspective. However, the perspectives imply

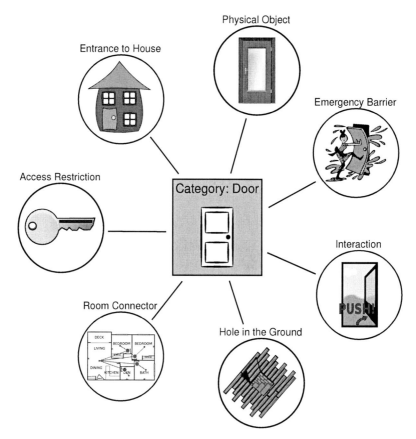

Figure 3.1: **Example for Different Perspectives on Spatial Information.** A Door category can be described from different spatial perspectives. An ontology module that complies with some of these perspectives specifies a Door by means of different properties and relations. (Images taken from http://openclipart.org, visited on November 15, 2011)

the requirements necessary to model the door category, i.e., they identify how a door is classified in the ontology module. The distinction between the different perspectives is nevertheless also crucial as two perspectives can potentially have contradicting definitions for the same category. For example, if one perspective describes trapdoors as holes in the ground and another perspective describes trapdoors as physical entities with masses and materials, the resulting inference that holes have masses and materials is normally incorrect.

For this purpose, we aim at keeping the distinction transparent between possible perspectives on space that an ontological module can comply with. Hence, the perspective indicates how an ontology module specifies respective spatial categories and their intended semantics. The perspectival distinction can also guide the development, selection, and combination of modules, e.g., if specific spatial perspectives are needed to solve some task or representation. However, selections and combinations have been pointed out as difficult issues:

> This surfeit of possibilities [for formalizations of space] should be an advantage, as we can pick rather precisely-tuned spatial descriptions as appropriate for the solution of various tasks. We might also beneficially ask questions concerning which aspects of the descriptions find empirical support as cognitively-plausible models and which do not. However current practice is somewhat different. The existence of alternative, perhaps competing, descriptions is in many cases placed on one side while a particular descriptive alternative is explored further. The problem of combining disparate descriptions remains a significant open issue. [Bateman et al., 2007a]

In this thesis, we aim at modeling these disparate descriptions as ontology modules that comply with different spatial perspectives. The perspectives guide the ontological formalization and indicate their intended meanings. They indicate requirements and characteristics of the ontology modules and they support combinations between modules with regard to the spatial perspectives they comply with. Also practically, a spatial system that relies on a combination of diverse spatial aspects benefits from the selection of ontology modules that are clearly separated by their thematically different perspectives. The approach is in contrast to integrating all spatial perspectives into one (monolithic) ontology module, which would not only need to deal with conflicting definitions (cf. the door example above) but also lead to a massively complex representation.

Furthermore, it is rather natural to have multiple perspectives on a domain expressed by different ontological representations: Although ontologies aim at specifying *consensual knowledge* of a certain domain [Gómez-Pérez et al., 2004], there is not automatically only one ontology that describes this domain in a correct or exhaustive way. In fact, there exist many ontologies of the same domain and they might even be incompatible with each other [Gómez-Pérez et al., 2004, p. 163][Euzenat and Shvaiko, 2007, p. 10]. This is caused either by different formalizations or thematically different perspectives:

- In the first case, content-identical information is formalized either by different logical axioms or by different design criteria. For instance, the specification for 'A is located at location B' may be specified by using a relation 'locatedAt' between two entities, possibly specified together with a time period, or it may be specified by using a category 'Location' that defines a 'located entity' and a 'location', i.e., reifying the relation 'locatedAt'. Although the intended meaning of this representation may be the same, its formalization differs and potentially the logic used too. The representations for 'locatedAt' and 'location' are thus not necessarily compatible with each other, unless a formalism provides a translation from one ontology specification to the other.

- In the second case, the purpose or emphasis of domain-specific information described differs. For instance, information about the domain 'plants' might be described by two ontologies, one specifying the biological and spatial structures of plants, the other specifying their impact on the environment. Both ontologies might either complement each other, i.e., the ontological specifications could be merged, or they might formalize incompatible types of information about the same domain, i.e., the axiomatization of the constraints of 'plant' in one ontology entirely differs from the definition in the other ontology.

This is typical for common-sense knowledge, because terms can vary across societies or contexts and change over time. Meanings are often not permanently fixed:

"A perspective is merely the result of someone's coming to cognitive grips with the world. Precisely because reality is so multi-faceted, we are forced to filter out some aspects of it from our attention which are less relevant to our purposes than others. Some of these processes of selection are performed deliberately and methodically. (...) Often, especially among scientists, the purpose of roping off a particular domain is simply to gain understanding of what the entities within it have in common, and of what makes them different from entities in other domains." [Munn and Smith, 2008, p. 11 f.]

Multiple perspectives not only stem from factoring in or out particular aspects for one domain. Also the definition of entities within one domain may vary according to different situations or contexts [Pike and Gahegan, 2007]. According to the context, one term can be defined in many ways, in which case different ontologies can model these different contexts best [Takeda et al., 1995]. Hence, different ontology modules can specify heterogeneous information about the same domain, possibly by using different logics. This heterogeneity can not only be analyzed for existing ontologies. It may also support the development of new ontologies. If an ontology module explicitly complies with one perspective to represent a domain, it consequently complies with formalizing respective categories and constraints from this perspective, and it can be compared with other ontology modules on the basis of their perspectives.

In the next section, we focus on the distinction of thematically different perspectives on space, which can be described by specific ontology modules. We also analyze how the perspectives indicate requirements and characteristics of their complying ontology modules.

3.2   Perspectives on Space

As space can be described and structured in different ways, different research areas have contributed to these aspects (see section 2.6). In particular, space can be seen from various perspectives to determine what needs to be defined about spatial information. For example, a perspective that originates from a spatial calculus only describes space with regard to certain geometrical objects and their calculus-based spatial relations. However, a perspective that originates from geographical information systems focuses on domain-specific entities, their spatial relations and attributes. To classify and distinguish the different perspectives on spatial information, we define them as follows:

Definition (Spatial Perspective). *A perspective provides the point of view from which certain aspects of a domain are described, i.e., it determines which aspects are part of the perspective and thus have to be taken into account. In particular, a spatial perspective SP determines which spatial aspects are characterized by it. Within a spatial perspective, (ontological) modules can be specified that comply with requirements of the aspects determined by the perspective. In particular, the perspective constrains the types of spatial categories and relations that are defined in such modules.*

Hence, a spatial calculus perspective can be distinguished from a geographical perspective, as they characterize spatial entities and relations from distinct points of view. An ontology that complies with one spatial perspective thus specifies respective spatial categories and relations. Based on the characterization of types of spatial information as discussed in the overview in section 2.6, this thesis distinguishes the following four different perspectives for spatial information:

1. quantitative and qualitative space,

2. abstract space,

3. domain-specific space,

4. multimodal space.

Each of these perspectives define space on the basis of certain criteria that determine which ontological categories can be specified for spatial information. The distinction between these four perspectives, qualitative/quantitative, abstract, domain-specific, and multimodal, is also motivated two-fold: first, modules that comply with one of these perspectives consequently have to follow specific design criteria and specification characteristics, e.g., they can model only ontologically high-level, abstract categories and relations or only low-level, specific categories and relations for spatial information; second, modules that comply with the perspectives are thematically separated, i.e., their contents and interpretations for spatial information provide different aspects and meanings. This results in an ontologically-driven characterization of space that leads to a comprehensive and modularly-structured description of space, and users or systems can select from this description, which spatial information to represent. [Hois, 2011]

The following questions can answer if an ontology module complies with one of the spatial perspectives: what categories are specified and how are they defined; what is their detail of axiomatization; how general or specific are the definitions; what is their level of granularity; what is the motivation behind the ontological definitions; which design aspects have been taken into account. This can be summarized by the type of the ontology module and by constraints over the modules' categories and relations. Table 3.1 illustrates the four different spatial perspectives used in this thesis and how they can be distinguished according to ontological and thematic aspects, i.e., according to the ontology types, categories and relations.

Table 3.1: Spatial Perspectives. An ontology module that complies with one of the spatial perspectives is defined by its ontology type and its ontological specification of categories and relations.

Spatial Perspectives	Ontology Type	Ontological Categories and Relations
Qualitative and Quantitative	Core, Domain	basic categories and relations, model- or data-grounded
Abstract	Upper-Level, Foundational	particulars, coarse granularity, high axiomatization
Domain-Specific	Task, Domain, Application	content-specific, fine-grained, specialized
Multimodal	Core	modality-dependent, empirically grounded

As a qualitative/quantitative perspective provides primarily basic spatial information with regard to formal mathematical models or data formats, it has to be specified as a core or domain ontology that defines the respective categories and relations for the basic spatial types. An abstract perspective provides high-level ontological classifications for general spatial information, and thus it has to be specified as a foundational or upper-level ontology that defines abstract spatial categories and relations. A domain-specific perspective focuses on spatial information used in particular domains or applications, and it is therefore specified as a task, domain, or application ontology that defines the domain-specific contents. Finally, the multimodal perspective models spatial information with regard to a specific modality, it is consequently modeled as a core ontology that provides modality-specific categories and relations. A detailed explanation for these different perspectives and their distinction between each other is presented in the following sections.

In essence, we classify ontology modules in terms of their perspective on the domain of spatial information. The perspectives and accordingly their modules separate the domain thematically into ontologically distinct groups, which are determined by their spatial aspects [Hois, 2010a]. This modeling is also motivated by information hiding and encapsulation realizing a separation of concerns. It also reduces complexity and results in clearly structured ontological specifications that are thematically identifiable. Applications or users can select or develop ontology modules for their needs on the basis of the perspectival distinction. However, as ontology modules that comply with different perspectives still refer to the same domain even by defining different aspects of it, they can be combined with each other. This process is discussed in detail in section 5.1.

Table 3.2 shows examples of ontology modules that provide specifications for spatial perspectives. An ontology module that complies with one of the spatial perspectives commits to their respective ontological characterizations. The following sections individually present in more detail the constraints of ontology modules for the fours perspectives, namely their ontological characteristics, requirements, and conceptual interpretations. The consequences for ontology modules that comply with the perspectives are exemplified.

Table 3.2: Spatial Perspectives Examples. Example ontology modules complying with different spatial perspectives.

Spatial Perspectives	Contents of Module Examples Complying with the Perspectives
Qualitative/Quantitative	Orientation, Distance, Polygons, Geometry
Abstract	Conceptual Spaces, Spatial Actions, Spatial Design Patterns
Domain-Specific	Geography, Astronomy, Architecture, Tourism
Multimodal	Gesture, Vision, Language

3.2.1 Quantitative and Qualitative Space

A quantitative or qualitative perspective on space represents spatial information on the basis of primitive spatial entities. These are conceptually limited and constrained to the intended specification and domain-specific representation. Consequently, an ontology module complying with the quantitative or qualitative perspective is either a core or a domain ontology (see section 2.6.3). As a core ontology, it defines basic categories and relations, e.g., jointly exhaustive and pairwise disjoint categories, which are sufficient to describe the types of spatial information in the ontology module. It typically axiomatizes the categories and relations in detail, i.e., properties of and relations between categories are specified exhaustively. As a domain ontology, the ontology module defines primitive categories and relations of spatial information with an emphasis on a specific domain, i.e., the ontology module specification is typically broader and less axiomatized though providing a detailed taxonomy.

Modules with quantitative or qualitative perspectives can be based on spatial (formal) models or data formats. For example, an ontology module that models the formalization of a spatial calculus complies with a quantitative perspective. As concrete spatial calculi analyze space in an axiomatic and rather abstract way, they do not define terminological aspects of space but abstract entities, such as points, lines, or polygons. For instance, if an ontology module specifies the region-based calculus RCC-8 (cf. section 2.6.2), it defines the primitive type for region, the region-based base relations, and the compositional constraints for combining base relations. As it therefore provides a specification for the primitive type, the exhaustive relations, and a high axiomatization based on composition, an ontology module for RCC-8 is consequently classifiable as a core ontology module. For example, a specification that re-formulates RCC-8 as an ontology by using topological representations has been presented in [Haarslev et al., 1998].

An example of an ontology module that complies with the quantitative perspective is the specification of geometrical objects inspired by the scalable vector graphics format (SVG). Such a specification defines two-dimensional, geometrical primitives on the basis of points, lines, curves, and shapes. For instance, the geometrical primitive of a circle can be defined by its center point, radius, and line type. As an SVG module is intended to specify the domain of geographical primitive entities and as it is less axiomatized in terms of properties and relations of categories, such an ontology module can be used for the specification of a respective domain ontology [Ipfelkofer et al., 2006].

In general, ontology modules with a quantitative or qualitative perspective specify the semantics of their intended types of spatial information based on mathematical models or spatial data formats. Hence, they are conceptually not comparable with the abstract, domain-specific, or multimodal perspectives. Nevertheless, the quantitative and qualitative perspective can provide a characterization of basic spatial entities that can be related with categories and relations from modules complying with other spatial perspectives.

Thematically, the categorization of qualitative and quantitative spatial information can be distinguished from the other perspectives as it is primarily concerned with categories that (qualitatively or quantitatively) describe volumes, shapes, sizes, distances, and measurement scales [Hayes, 1985]. Hence, ontology modules complying with the qualitative or quantitative perspective typically specify these types of categories. They may also introduce ordering relations, dimensionality, unit intervals, or measure functions. Although these properties can be applicable for modeling domain-specific categories, they can clearly be separated from concrete domains, i.e., from a domain-specific perspective, and modeled individually by quantitative or qualitative perspectives. More important, this specification of purely quantitative or qualitative categories and relations in an ontology module can be reused by other modules from different perspectives, in particular, from the domain-specific perspective.

In terms of applicability, quantitative and qualitative spatial modules can be used directly or in combination with other modules by systems and users. For example, in the geographical domain both quantitative and qualitative types of spatial information are needed to model topographical information: an ontology for this has to specify types of landforms that are defined by shapes, i.e., it needs to specify qualitative geometry, e.g., convex, concave, and cone-shaped properties, as well as the quantitative cartographic data [Mark and Smith, 2004].

The evaluation of ontology modules complying with quantitative and qualitative perspectives depends on their contents and use. If ontology modules are based on quantitative data models or qualitative axiomatic calculi, their correctness is (already) proven by the external spatial format that is modeled by the modules. Such proofs may also be integrated into the ontological module, if the ontology language is expressive enough. If quantitative or qualitative spatial modules are reused in particular domains, their applicability can be evaluated in a data-based way, i.e., test sets with correct (expected) classifications can verify the accuracy of the ontological representation and classification. In general, however, consistency and satisfiability of ontology modules as well as correctness of their axiomatic constraints or proofs can be analyzed by the respective ontological reasoners.

In summary, modules complying with qualitative or quantitative perspectives are specified as core or domain ontologies and they define basic categories or relations motivated by spatial models or based on data formats. With regard to these spatial models and data formats, the ontology modules provide a fine-grained specification or axiomatization for spatial information. Specific ontology modules that comply with the quantitative or qualitative spatial perspective are introduced in section 4.2.1. In particular, we present a qualitative ontology module that reflects region-based spatial information on the basis of RCC-8 and we present a quantitative module that reflects constructional spatial information based on an architectural structure format.

3.2.2 Abstract Space

An abstract perspective on space represents spatial information on a fundamental and high level of abstraction. Ontology modules that comply with an abstract spatial perspective specify categories and relations regardless of any domain, task, or system. They are not based on external data models, and they are typically not developed semi-automatically, e.g., by using machine learning techniques with domain-specific data sets. In contrast, their specifications are created manually by experts or ontology designers. The specifications can be inspired by insights and methodologies from other disciplines, such as philosophy, mathematics, psychology, or cognitive science. As a consequence, ontology modules complying with an abstract spatial perspective can be classified as foundational or upper-level ontologies (see section 2.6.3).

As modules with an abstract space perspective provide high-level information on space, their categories and relations cannot have direct instances. For example, in the case of DOLCE the category Spatial Location is subsumed under Physical Quality, which provides a too general description to have instances. The category would need to be reused and refined by further subcategories by some non-foundational ontology modules, which in turn can have instances. Hence, although Spatial Location cannot have direct instances, its subcategories may.

As the categories and relations of an abstract perspective provide foundational characterizations of space that are domain-independent, their detail of specification has a coarse granularity. Nevertheless, as required of upper-level or foundational ontology their specifications provide a high axiomatization for their entities. Consequently, modules with an abstract space perspective model the relations and their constraints on categories as exhaustively as possible. High axiomatization, however, may conflict with or may not be supported by the ontology language, in which the module is formulated. For example, it may be necessary to reduce axiomatization to provide a logically decidable formalization if this is technically necessary. Hence, a module with an abstract space perspective may be designed with a low degree of axiomatization caused by modeling decisions.

An example of an ontology module complying with the abstract spatial perspective is a module that specifies image schemata [Johnson, 1987], which were introduced in section 2.6.1. Such an ontology module is clearly inspired by results from psychology and cognitive science, however, not directly based on external data formats or formal specifications that are used by quantitative/qualitative modules. An image schemata module provides information about the basic recurring spatial patterns, e.g., it specifies the categories CONTAINER, SURFACE, LINK, PATH, NEAR-FAR, PART-WHOLE, and CENTER-PERIPHERY. The respective relations and constraints of these categories are specified by the image schemata module. A computational formalization that implements image schema types has been presented in Kuhn [2002], in which these abstract types have primarily been reused to model geographical entities and their instances.

Compared with the other spatial perspectives, the abstract perspective provides the most independent and general representation of space. Compared with quantitative/qualitative perspectives, it is not bound to external data formats or mathematical models, and compared with domain-specific or multimodal perspectives, it is not constrained to some special domains or modalities. On the contrary, the abstract perspective aims at a comprehensive yet high-level spatial characterization. However, as it provides the most abstract representation, modules complying with this perspective can be reused by other modules from different perspectives. In particular, modules complying with domain-specific spatial perspectives can often apply and refine the specifications and modeling decisions from the abstract perspective [Hois, 2011].

Because of the task- and domain-independence of the abstract spatial perspective, its modules are difficult to evaluate as there is no formal reference system to compare to. Although their cognitive adequacy can thus only be judged by peer-reviews, abstract spatial modules can be tested by means of their reusability and interoperability. To test reusability, domain-specific ontology modules that reuse an abstract spatial module in order to refine categories and relations for specific domains or tasks can be evaluated with regard to their task performance or classification results. As these domain-specific modules are based on the general structure and characterization of abstract modules, the abstract specifications are evaluated indirectly. To test interoperability, the abstract module can be analyzed with regard to its reference potential, i.e., if the abstract module can support integrations of different modules that reuse the same abstract module. Here again, the technique of ontological modularity provides a practical mechanism to relate the abstract module and its refinements, which is discussed in more detail in section 5.1. In general, modules complying with an abstract perspective on space form a central component to provide an understanding of spatial structures and their relations and they can guide the development of more specific, fine-grained ontology modules. They can provide spatial patterns and design characteristics for spatial entities for other modules, which can apply the general distinctions of spatial categories and relations. If different modules reuse the same abstract space module, they may also be compared or combined more easily.

In summary, modules complying with an abstract spatial perspective are classified as foundational or upper-level ontologies, they define spatial categories with a coarse, high-level granularity, and they provide a high axiomatization if needed. They can be motivated or inspired by other disciplines that analyze space, and they can provide ontological formalizations for these theories or methodologies. Abstract modules can be evaluated by means of peer-reviews, reusability, and interoperability. Specific ontology modules that comply with the abstract spatial perspective are introduced in section 4.2.2.1, where we present an ontology module for spatial actions that is reused by domain-specific spatial modules.

3.2.3 Domain-Specific Space

A domain-specific perspective on space represents spatial information on the basis of particular tasks, applications, or systems. Spatial entities and their properties are accordingly characterized from a specialized and content-specific point of view. Ontology modules that comply with the domain-specific perspective formalize respective entities in combination with an application. They provide *terminological* aspects to specify particular characteristics of space, e.g., a specific application-driven spatial topic. Hence, these modules can be classified as domain ontologies, which are also called task ontologies or application ontologies depending on their purpose (see section 2.6.3). Typically these modules provide a fine-grained characterization of their domain-specific categories and relations. However, the level of granularity is also determined by application or task purposes.

As spatial domain-specific modules are closely related to the systems they are developed for, their categories and relations are often available as system-specific representations or external data sources. These representations or sources reflect the instances that are supposed to be modeled by the domain-specific modules. Consequently, they provide a reference frame that guides the development of the modules. Also, they ground the ontological representation by providing actual data sets, and they determine the requirements for the module specification. As the modules are designed for particular tasks or applications, characterizations of spatial information from different ontology modules that comply with the domain-specific spatial perspective can vary highly and even become incompatible with each other.

Clearly, the number of domains is infinite and there exist endless possibilities of domain-specific spatial modules. According to Lefebvre [1991, p. 33] three domain-specific types of space can be distinguished: (i) 'spatial practices', i.e., named entities in the environment, such as parks, fields, museums; (ii) 'representations of space', i.e., conceptual depictions of space, such as maps and plans; and (iii) 'spaces of representation', i.e., the sum of the two previous types, such as shopping malls, which are termed social spaces. Even though this distinction does not provide a precise ontological categorization of spatial domain-specific types, it highlights possible overlaps and conflicts of different spatial domains. Hence, Lefebvre [1991, p. 86] points out that the spatial information we are confronted with is an "unlimited multiplicity or uncountable set of social spaces" with different granularity and interpretations.

An example of an ontology module that complies with the domain-specific spatial perspective is a module for architecture. The entities and relations it specifies are determined by the architectural domain or by a specific application for architectural tasks, e.g., the designing of a building. However, even for this domain-specific architectural perspective, different ontology modules can be designed with regard to specific architectural interpretations, as architectural designs can be seen in many different ways. For instance, buildings can "be viewed as works of art, as technical achievements, as the wallpaper of urban space and as behavioural and cultural phenomena" [Lawson, 2001, p. 4].

Furthermore, such an architecture module can reuse existing spatial characterizations, possibly provided by abstract spatial ontology modules. By using them, the architecture module can benefit from an existing ontological structure of the domain in general, and it can also reuse abstract spatial patterns and classifications. For example, a category for buildings in the architecture module can regard this entity as a container for other entities. A similar reuse strategy has been used in Kuhn [2002] to model houses in the geographical domain, as discussed in the previous section.

Compared with the other perspectives, the domain-specific perspective on space stands out as being customized for specific applications, tasks, or purposes. Modules with other spatial perspectives provide a more basic or high-level specification of spatial categories. Although these can be reused by domain-specific modules, the domain-specific perspective itself does not provide such abstract, basic, or high-level ontological characterizations of spatial information. For example, a domain-specific ontology module that is applied in a human-computer interaction system may reuse modules with a multimodal perspective. However, domain-specific ontology modules can also be combined or aligned with each other, as they can describe the same specific domain by modeling different fine-grained aspects of it.

As modules complying with the domain-specific perspective provide a representation for the needs of a spatial system, application, or task, their applicability is analyzed with regard to these needs. Hence, the modules are evaluated in terms of their use in applications or with regard to external domain sources. In particular, the results that an ontology module, its classification, or reasoning achieves within the system are analyzed with regard to their performance or correctness. For example, a gold standard for classification results may be provided by the system, which can then be used to analyze whether the ontology achieves the same classification results. Alternatively, a system can perform its tasks with and without the ontology module for the domain, and the overall performance results can be compared.

In summary, modules complying with a domain-specific perspective provide a representation for spatial information that is determined by specific systems or applications. Their specifications are goal-oriented or task-driven and typically refer to an external domain-specific spatial source within the applications that use the modules. Concrete types of categories and relations for spatial information can highly vary across modules with a domain-specific perspective. Similarly, their detail of axiomatization and level of granularity is determined by the requirements of the systems or tasks. Specific ontology modules that comply with the domain-specific spatial perspective are introduced in section 4.2.3. In particular, we present domain-specific modules for physical entities in image recognition systems, for architectural entities in architectural design tasks, and for smart home entities in home automation systems.

3.2.4 Multimodal Space

A multimodal perspective on space represents spatial information on the basis of a specific sensory modality. Examples for multimodal perspectives are vision, audition, or tactition. Spatial entities and their properties are described from these points of view by the modal perspective. Ontology modules complying with a specific multimodal perspective formalize these types of spatial categories and relations. As such ontology modules are limited to a narrow classification for spatial information, they can be classified as core ontologies (see section 2.6.3). The way they characterize space is solely based on the modality they reflect.

Consequently, module specifications for a multimodal perspective are empirically grounded in modality-related models or data. The categories and relations they specify are fine-grained and they have a high axiomatization with regard to the modality type. Also, multimodal modules differ from conceptual ontology modules, as they are restricted to multimodal resources. If an ontology is developed on the basis of a modality, its resulting classification is by no means influenced by general commonsense knowledge about space. It only distinguishes ontological categories in terms of their modality-dependent differences. Also, the axiomatization and inference results in multimodal modules can differ from conceptual modules.

For example, mereological inferences about the part-of relation do not necessarily hold for ontological specifications that are based on a linguistic modality. An example in Lang [1991] demonstrates that the transitive relation being-a-component-of between two categories is not transitive from a linguistic perspective, e.g., for the following set of categories: fridge, kitchen, flat, house. Although a fridge is a component of a kitchen, a kitchen is a component of a flat, and a flat is a component of a house, a fridge is not necessarily a component of the house. The transitivity constraint for the part-of relation is violated, i.e., the part-of relation is not defined as transitive from a linguistic spatial perspective.

Another example of an ontology module complying with a multimodal perspective is a spatial module for vision. Such a vision module accordingly reflects spatial information only on the basis of visual features. For instance, such a module can specify categories for texture or colors and define these visual properties. Therefore, the module may apply insights from human perceptual recognition of texture and colors, or it may reuse standardized color models that are based on visual perception. The ontological formalization for a visual-spatial representation that has been introduced in [Maillot and Thonnat, 2008] provides texture-related categories inferred from experimental results in cognitive science and color-related categories derived from a standardized color dictionary.

Compared with the other perspectives, multimodal modules are thus determined by their modality-specific description of space. Compared with the quantitative/qualitative perspective, the multimodal perspective is not based on metrical spatial categories or formal models but inspired by cognitive perception of space. Compared with the abstract perspective, it is not bound to high-level categories but defines spatial information on a fine-grained level. And compared with the domain-specific perspective, it is not designed to satisfy application- or task-specific requirements.[1] However, as already discussed in the previous sections, modules with a multimodal perspective can be related to modules from other perspectives.

---

[1]This also means that ontological modules designed for specific applications are often domain-specific. For example, the ontological module for vision recognition presented in section 4.2.3.1 is not a multimodal (visual) ontology about visual features but a domain-specific module for object categories that can be visually perceived.

Indeed, as the different modalities are channels for humans to perceive and categorize space, they implicitly influence other spatial categorizations that are designed by humans:

> In order to understand our relationship with space, we first need to explore how we become aware of it. Primarily of course we see it, since it is largely evident to us visually. (. . .) it is easy to forget that space is also perceived through the sensations of sound, smell and even touch. Perception is actually more than just sensation. Perception is an active process through which we make sense of the world around us. To do this of course we rely upon sensation, but we normally integrate the experience of all our senses without conscious analysis. [Lawson, 2001, p. 42]

For instance, the spatial aspects of the foundational ontology DOLCE are inspired by cognitive and perceptual empirical results how humans perceive the world and its entities. Here, however, we separate the different modalities into modules to clarify their individual contributions to spatial information.

Although the multimodal perspective does not depend on any specific application, it can nevertheless be useful in spatial systems, in particular, in contexts of agent-oriented systems that follow an embodied or action-oriented view of their domain [Gangemi, 2010, p. 146] or in contexts of human-computer interaction systems [Kruijff et al., 2007]. For example, an ontology module with a linguistic perspective on space can provide a categorization that models how humans talk about spatial situations and processes, and its development can be based on linguistic resources and empirical insights. Hence, the evaluation of multimodal ontology modules can be performed by using external, empirical data sources. In addition, their performance in applications or systems can be analyzed, if applicable.

In summary, modules complying with the multimodal perspective are classified as core ontologies as they provide a fine-grained specification on space solely based on one modality. The specification of their categories and relations is accordingly motivated by modality-based categorizations based on empirical data or cognitive insights. The modules provide a high detail of axiomatization and a high level of granularity for their spatial characterization. A specific module that complies with the multimodal spatial perspective is introduced in section 4.2.4, where we present a linguistic ontology.

3.3 Chapter Summary

In this chapter, we have presented a perspectival distinction for ontology modules to model the diversity prevalent in spatial information. The four perspectives (quantitative/qualitative, abstract, domain-specific, multimodal) provide different ways for describing space. They also determine the requirements of ontology modules that comply with specific perspectives. In particular, they determine the modules' ontology types, specifications of categories and relations, their detail of axiomatization, their level of granularity, their motivations and design decisions. Also, applicability, reusability, and evaluation strategies for the different perspectival modules were discussed.

Based on the distinction of spatial perspectives, modular ontologies for space have been developed for different perspectives, which are presented in the next chapter. To provide an exhaustive view of the domain of space, to connect and combine different perspectives, and to satisfy certain application- or task-specific requirements, we discuss extensions of the perspectival ontology module framework in chapter 5. Finally, applications and evaluations are presented in chapter 6.

4

Modular Ontologies for Spatial Information

In the previous chapter, the perspectives that distinguish different types of spatial information were conceptually identified. In this chapter, concrete spatial ontologies are developed and presented that comply with the spatial perspectives described above. In order to take into account these perspectives on spatial information, the ontologies are developed in a modular way, each module reflecting spatial information from one perspective of space. The general approach for developing such ontologies is described in this chapter and the resulting modular ontologies for space are presented. Also, the kinds of relations and combinations relevant among the spatial modules are analyzed, and their formalizations are presented in chapter 5.

This chapter starts with a general overview of the spatial ontology modules that are developed in compliance with the spatial perspectives framework given in section 3.2. Development and design guidelines as well as technical details are discussed. Furthermore, the use of a modular approach for developing the spatial ontologies is motivated with regard to technical as well as content-based aspects. The spatial ontologies developed in this thesis are then presented individually for each spatial perspective. Their respective purposes, requirements, and development processes are explained. Use and application of the spatial ontology modules are summarized, and combinations of different modules are outlined.

4.1   Overall Design of Modular Ontologies for Space

The previous chapter has discussed how semantically different perspectives can be distinguished, which all describe certain aspects of spatial information. These perspectives may complement or partially contradict each other or point out entirely dissimilar aspects of spatial information, however, they all have in common the same domain, namely space. For the spatial ontology modules, we aim at an ontological representation that appropriately reflects such spatial perspectives and that adequately matches their semantic distinctions.

For this aim, the ontological representation has to satisfy the following requirements: As the perspectives are clearly distinguishable, their ontological specifications should also be distinguishable. For example, if an application or a user wants to access only information from one perspective (cf. table 3.2 for examples), the respective ontological specification should be available independently from other perspectives, i.e., the specification should be available as an isolated ontology module, if possible. Consequently, the modules are available as individual ontological representations. As these representations share the same domain by formalizing different spatial information, they may have some parts in common that are related with each other. These common parts and relationships are ideally provided in an ontological representation as well. For example, if an application requires to reflect combinations of different perspectives and spatial aspects, the single representation should comply with some methods that allow a well-defined access to a combined representation of different single ontological representations, i.e., different perspectives.

As a result, we aim at a representation that provides the semantically distinct spatial aspects, each of which is specified as one ontological module. Each ontology module then complies with one of the spatial perspectives, described in chapter 3. At the same time, their combinations use different technical combination methods, described in chapter 5, to support applications that require several aspects about spatial information.[1]

---

[1] Several cognitive theories assume a modularly structured mind for competences that support separate cognitive domains. In particular, functionally specialized (domain-specific) parts are expected to work together with each other in order to produce behavior [Newcombe and Ratliff, 2007]. This direction is, however, not further investigated for the purpose of this work.

Hence, this ontological representation consists of single syntactically distinct ontology modules that reflect semantically distinct perspectives and it relates the modules with each other according to the spatial information they specify. The syntactic distinction allows a technically easy use and access and it reflects the semantic distinction according to their spatial contents. In particular, an application can be provided with the relevant parts of information for its purpose ignoring other information:

> Ideally, we want to represent only knowledge which we will need for solving the tasks which we have to solve and abstract from all other knowledge. Since we typically collect knowledge before we know the specific tasks for which we will use it we can not be so restrictive; but depending on the classes of tasks we want to solve we are usually able to exclude a range of items which we will never consider. [Freksa, 1991]

A modular ontology representation for spatial information can thus also be beneficial for applications as it provides them only with those parts that are relevant for their specific requirements or purposes. The applications thus have to understand and apply only small parts (modules) that reflect the information they need. Moreover, the representation directly supports the reuse of modules by flexible module combination techniques, it separates and encapsulates the different aspects of spatial information, and it reduces complexity by means of clearly structured specifications [Kutz et al., 2010a].

Figure 4.1 illustrates the different types of ontologies that are specified according to their spatial perspectives, namely foundational, domain-specific, qualitative-quantitative, and multimodal. They also provide a connection to external sources for spatial information (independent of ontological information): foundational ontologies provide abstract information about space that is independent from any application or use, terminological ontologies provide information about domain-specific spatial aspects, qualitative and quantitative spatial ontologies are closely related with formal modeling of space, and multimodal ontologies are closely related to human-computer interaction scenarios.

Concrete ontology modules that are designed according to this framework and that reflect certain spatial information are presented in section 4.2. The development process

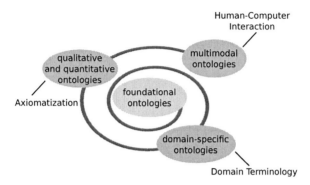

Figure 4.1: Spatial Ontology Module Types. Ontologies that reflect different types of information and that can be linked to external sources.

that underlies the modules ontologically (section 4.1.1) and technically (section 4.1.2) is discussed in the following.

4.1.1 Modeling Decisions

General modeling criteria for the development of spatial ontological specifications are primarily taken from approaches in formal ontology. As these are influenced by their philosophical roots in formal ontological research and information science, they provide ways for developing an ontology according to certain philosophical considerations about generally possible types and relations in ontologies. Jansen [2008b, p. 160 f.] summarizes eight characteristics any formal ontology should comply with:

- *Ontological Grounding*: Taxonomies should classify things on the basis of traits belonging to those things. An ontological category therefore has to classify objects by means of their characteristics and relations that exist in reality, i.e., a category definition is grounded in real things. Thus, spatial categories should be related to spatial objects and their characteristics and relations.

- *Structure*: Taxonomies should take into account subtypes. As pointed out above (see section 2.3 on ontology design and development), hierarchical dependencies are a crucial aspect in ontologies and necessarily specified by ontology modules.

If spatial entities are related hierarchically, the ontology should reflect this dependency. For instance, a spatial category 'WalkingEvent' could be subsumed by 'Event'.

- *Disjointness*: Types on the same level of classification should be disjoint. As far as possible, sibling categories should be specified as disjoint and provide a distribution of their supercategory. For example, 'WalkingEvent', 'DrivingEvent', 'FlyingEvent' can be subsumed as 'Event' and can be specified as mutually disjoint.

- *Exhaustiveness*: Taxonomies should subsume all the entities they are supposed to classify. Such a closed-world assumption is, however, often not possible and not the primary aim of an ontological specification. However, spatial ontologies developed for specific spatial applications should specify the categories that are required by the applications.

- *No Ambiguity*: Taxonomies should use terms unambiguously. Clearly, a category 'Event', for instance, should not appear twice in an ontology module with different names or formalizations.

- *Uniformity*: Taxonomies should have a well-defined domain. This question of granularity and consistency of module contents is particularly addressed by the spatially perspectival frame that provides a well-defined domain for each ontology module.

- *Explicitness and Precision*: Taxonomies should describe types in an explicit and precise way. The specifications of categories therefore need to be well-defined and distinguishable from each other. For instance, the spatial ontology modules provide as much axiomatization as necessary about spatial information they specify, and they should also follow the perspective-driven formalization for distinguishing the different types of spatial information because this directly supports the required precision.

- *No Meta-Types*: Taxonomies should not contain information about themselves, i.e., meta-categories, which specify aspects about other categorizations, should not be formulated in the same ontology module as the categorizations they specify. The modularized representation for spatial ontologies, however, cannot have meta-types: Each module complies with one spatial perspective on space, i.e., they cannot have retrospective specifications.[2]

These philosophically-inspired recommendations are applied for developing the spatial ontology modules. They provide a general characterization how the categories of a module should be specified. These criteria are used as general design guidelines, assuring that ontology categorizations are appropriately and concisely reflected. Furthermore, philosophical studies may help to decide on main distinctions in taxonomies. For example, the following basic categories provide a top-level classification on the basis of the Aristotelian categories (cf., e.g., Neuhaus et al. [2004]; Jansen [2008a]):

- *dependent* (accidents) and *independent* (substances) entities, i.e., the latter type does not need entities in order to exist and the first type requires entities from the latter in order to exist. For example, a forward-motion cannot exist without an entity that actually carries out this motion (or is forced to) but the existence of a spatial entity itself does not depend on any other entities.

- *continuants* and *occurrents*, i.e., entities that exist over time without changing and entities that occur in time. Hence, spatial objects that exist in space can be distinguished from spatial events that occur over a certain timespan.

- *universals* and *particulars*, i.e., entities that can be either general or individual. In particular, universals provide the basis for classifying particulars, and every particular is an instance of one or more universals. The distinction between universals and particulars, for example, is made by the distinction between the ontological specification of a taxonomy (TBox) and its instantiations (ABox), which is used for the spatial ontology modules.

- *complex* entities, such as states of affairs, sets, mereological sums, and classes. If combinations of spatial categories are required, for instance, a spatial plan for a travel route, then a complex category is defined.

---

[2]In contrast to meta-types, meta-information about the spatial ontology modules is specified for each module and formulated as ontology annotations.

These general distinctions should commonly be applied when formalizing ontological specifications [Masolo et al., 2003]. The spatial ontology modules, however, not always use exactly these categories (and category names). They, however, comply with these general distinctions and use them as an underlying structure.

Furthermore, the OntoClean methodology [Guarino and Welty, 2002] provides a tool to analyze an ontology with regard to similar distinctions between rigidity, identity, unity, and dependence (see section 2.3.3). For example, an ontology can be developed according to OntoClean's definitions, which has been applied for some of the spatial ontology modules (see below).

Orthogonal to these categorization distinctions for ontologies, their development process addresses the following criteria, as already outlined in section 2.3.2: Every module is required to be developed according to the spatial perspectives framework, i.e., it either formalizes aspects for only one spatial perspective or it formalizes combinations of clearly distinct perspectives. Every module has to provide meta-information about the categories and relations that it specifies, and the requirements an instantiation of it has to satisfy. Furthermore, every module has to describe its purposes and aims for which the ontological specification is designed. Therefore, it also gives information about the requirements it has to satisfy according to purposes and aims. Potential reusable ontologies or similar representations that guide the development process of the ontology module are analyzed. And where applicable, spatial systems and application scenarios are presented that use the spatial ontology module.

4.1.2 Technical Details

The spatial ontologies developed below are primarily meant to be used within spatial applications. Thus, from a technical point of view, they need to be specified in a way that can be accessed conveniently by the spatial applications. As surveyed in section 2.3.1, DLs are not only widely used and a common standard for ontology specifications but they also provide constructions that are general enough for specifying complex ontologies [Horrocks et al., 2006]. Moreover, they provide a balance between expressive power and computational complexity in terms of reasoning practicability. Thus, DL ontologies are used for the spatial ontological modules as far as possible. To facilitate the use of DL reasoners directly for the spatial applications, the ontologies are formulated in OWL (cf. section 2.3.1 on logical formalizations). Although the ontology definition on page 19 can be used for the specifications of the ontology modules, all of them have to specify the spatial perspective SP they comply with, which is one of $\{qualitative\text{-}quantitative, abstract, domain\text{-}specific, multimodal\}$.

Examples of the spatial ontology modules below are formulated in OWL's Manchester syntax [Horridge and Patel-Schneider, 2008]. Listings 4.1 and 4.2 explain the constructors of this syntax for categories and properties respectively, which are used below. As can be seen in the listings, category names are capitalized and relation names are not capitalized, which follows the naming conventions for OWL ontologies[3]. If necessary, the namespace is given in front of the category or property name, e.g., PhysicalEntity:BuildingType. Following these standards provides a technically well supported integration into the spatial applications (see chapter 6 on different use cases for spatial ontology modules). However, in case more expressive languages than DLs are needed and used, these are introduced and explained together with the application whenever necessary.

Listing 4.1: Overview of OWL constructions for category definitions (in Manchester OWL Syntax)

```
Class: CategoryName (followed by its specification)

    Annotations:
        rdfs:comment   ' General comments (and meta-information)
        about the category CategoryName. ' @en

    SubClassOf:
    rdfs:comment   ' Class construction of CategoryName ' @en
        SuperClassCategoryName ,
    rdfs:comment   ' Example of number restriction of a data property
    and its data-type range. ' @en
        dataPropertyName exactly 1 data-type ,
    rdfs:comment   ' Example of an existential restriction of an
    object property and its related category. ' @en
        objectPropertyName some AnotherCategoryName

    DisjointWith:
    rdfs:comment   ' Disjointness specification with another category ' @en
        YetAnotherCategoryName
```

---

[3]http://www.w3.org/TR/owl2-overview

Listing 4.2: Overview of OWL constructions for category definitions (in Manchester OWL Syntax)

```
ObjectProperty: objectPropertyName

    Annotations:
        rdfs:comment    ' General  comments  (meta−information )
        about  the  property  objectPropertyName. ' @en

    Domain:
    rdfs:comment    ' Specification  of  categories  that  are  able  to  have
    this  property ... ' @en
        CategoryNameOfObjectPropertyDomain                      .

    Range:
    rdfs:comment    ' ... and  the  range  the  property  can  have. ' @en
        CategoryNameOfObjectPropertyRange

    SubPropertyOf:
    rdfs:comment    ' Hierarchical  information  of  properties
    if  available. ' @en
        superObjectPropertyName

    InverseOf:
    rdfs:comment    ' Inverse  properties  if  available. ' @en
        CategoryNameOfInverseObjectProperty
```

To support the accessibility of the spatial ontology modules, all of them are available online[4]. This is not only common practice for OWL ontologies in general but it also allows spatial systems to access the spatial ontology modules directly. This, however, also requires the modules to provide enough meta-information about their potential usage. The combinations of the modules are also available as far as they are required by spatial systems, described in chapter 6. Section 5.1 shows the different ways of combining the modules and their specification for the accessible spatial ontologies.

In summary, every ontology module has to provide detailed information about the following aspects on the basis of the modeling decisions and technical considerations: The spatial perspective of the module, definitions of its concepts and properties (relations) and their axiomatization, how it can be instantiated or whether it can be grounded in external data, the module's purposes, aims, requirements, possible ap-

---

[4]http://www.informatik.uni-bremen.de/~joana/ontology/SpatialOntologies.html

plication examples, specific aspects of its development process, its expressivity and availability.

4.2 Ontology Modules for Different Spatial Perspectives

In this section, the ontology modules developed for the different spatial perspectives are introduced and technically explained. They follow the perspective distinctions made by the spatial categorization presented in chapter 3. The modules are used by different spatial applications presented in chapter 6.

4.2.1 Ontology Modules for Quantitative and Qualitative Space

Ontology modules that follow a quantitative or qualitative perspective are spatial specifications that are either closely related to spatial data formats or qualitative spatial models. Such information is relevant for a variety of spatial applications, for example, in the areas of architectural design, GIS, or robotics, in which spatial environments are represented that are strongly connected to spatial data. Thus ontology modules can often be grounded in such data. Also, spatial applications may use collections or combinations of quantitative and qualitative specifications to support different requirements [Bhatt et al., 2012].

4.2.1.1 Region-Based Ontology

The Region-Based Ontology (RBO) is a module that specifies ontological relations on the basis of the relations defined by the RCC-8 qualitative calculus (cf. section 2.6.2). The module's primary aim is to provide a region-based set of relations among entities according to the RCC-8 relations, which can be reused by other modules to formulate region-based qualitative spatial relations among their domain-specific entities. The requirement of the RBO module therefore is to transfer RCC-8's topological relations to ontological relations in the module. As the RBO module's purpose is to provide such an ontological RCC-8 specification, it is not directly related to specific applications. Hence, no particular application requirements affect the specification of the RBO module. Other spatial modules that are introduced below (e.g., the ACO and ADO modules in sections 4.2.1.2 and 4.2.3.2), however, reuse the RBO module to support certain applications and domains.

In general, RCC-8 as a qualitative logical representation is often specified by a topological interpretation of the modal logic **S4** [van Benthem and Bezhanishvili, 2007] that encodes the RCC-8 relationships. For this purpose, the modal operators are interpreted topologically, namely the necessity operator $\Box$ is interpreted as the interior operator **I** and the possibility operator $\Diamond$ as the closure operator **C**. Sentences are then built by using the modal operators together with propositional variables [Bennett, 1996].

To provide the technical support for reusing the *RBO* module in other OWL modules, the ontology module needs to be specified in OWL as well. This way other modules can directly import the *RBO* module and use its relationships [Hois et al., 2009a]. Although the specification of RCC-8 in DL is not expressive enough to formalize the full composition table of the qualitative calculus [Haarslev et al., 1998], it formulates the terminology of RCC-8's relations and their ontological properties. For example, the reflexivity and transitivity properties of RCC-8 relations are specified in the *RBO* module. And as full composition tables are often not required for application purposes owing to incomplete topological knowledge [El-Geresy, 1997], specific spatial reasoning can be used that provides an approximation of RCC-8 composition table dependencies, e.g., supported by the DL reasoner RacerPro [Racer Systems, 2007], as described in section 6.2.[5]

As the *RBO* module describes a qualitative perspective on space it only defines a core set of entities and relations, namely the most-general type Region (specified as a subtype of Thing), and the RCC-8-related relations (ObjectProperties) of the category Region. Consequently, a class hierarchy is not further specified by this module, because its relations are not defined for specific category types except for regions.

Based upon the relation naming in RCC-8, the *RBO* module defines the following relations: disconnectedFrom (disconnected), externallyConnectedTo (externally connected), partiallyOverlaps (partial overlap), equalTo (equal), tangentialProperPartOf (tangential proper-part), nonTangentialProperPartOf (non-tangential proper-part), inverseTangentialProperPartOf (tangential proper-part inverse), and inverseNonTangentialProperPartOf (non-tangential proper-part inverse) (cf. section 2.6.2). In addition to these 8 basic relations,

---

[5]Here, the RCC-8 specification is primarily formulated in OWL for technical practicability. However, if computational complexity and technical applicability is less important, e.g., in cases where only the qualitative representation of a calculus is relevant, other representations for spatial logics can be used. In Hois and Kutz [2008b], for example, a qualitative spatial calculus is related with another ontology module formulated in CASL [Mossakowski et al., 2008].

the *RBO* module also introduces further relations that subsume the basic relations in order to construct a property hierarchy in the ontology that illustrates the dependencies between RCC-8 relations [Cohn et al., 1997a]. For example, the relations disconnectedFrom and externallyConnectedTo are subsumed by the relation discreteFrom, and the relations tangentialProperPartOf and nonTangentialProperPartOf are subsumed by the relation properPartOf. These subsumptions match with the relationship between RCC-8 and RCC-5: when mapping the basic relations from RCC-8 to RCC-5 the subsumed relations coincide.

The ontological specification of the RCC-8 relations extends the OWL specification that has been introduced by Grütter et al. [2008]. The extension in the *RBO* module provides more details and refinements in the ontological specification. In particular, characteristics of the RCC-8 relations, such as reflexivity, symmetry, transitivity, and disjointness, are newly introduced. For example, nonTangentialProperPartOf and inverseNonTangentialProperPartOf are specified as transitive, discreteFrom as irreflexive, and equalTo as reflexive.

Listing 4.3 shows the specification of the ObjectProperty equalTo in the *RBO* module. As for all 8 basic relations, the OWL annotation comment indicates the relation name in the RCC-8 terminology. The ontological characteristics of the ObjectProperty equalTo are symmetric, transitive, and reflexive, and disjoint with the other 7 basic RCC-8 relations. Together with proper-part and partially overlap relations, equalTo is subsumed by the ObjectProperty overlaps.

Listing 4.3: Specification of the property equalTo in the *RBO* module

```
ObjectProperty:  equalTo

    Annotations:
        rdfs:comment   ' RCC−8  relation   'equal '  ' @en

    Characteristics:
        Symmetric,
        Transitive ,
        Reflexive

    SubPropertyOf:
        overlaps
```

DisjointWith :
rdfs : comment   ' Disjoint with the other 7 basic relations in RCC–8 ' @en
    disconnectedFrom ,
    externallyConnectedTo ,
    inverseNonTangentialProperPartOf ,
    inverseTangentialProperPartOf ,
    nonTangentialProperPartOf ,
    partiallyOverlaps ,
    tangentialProperPartOf

As discussed in the overview of RCC, the RCC-8 relations build a jointly exhaustive and pairwise disjoint (JEPD) set of basic relations. Not only equalTo but also the other 7 basic relations are disjoint with each other, i.e., they satisfy the definition of 8 pairwise disjoint relations in RCC-8. The requirement that the 8 basic relations be jointly exhaustive in RCC-8 is only implicitly given by the *RBO* module that does not specify further properties, except for the additional superproperties of the 8 basic relations introduced in the module to organize the basic relations. Figure 4.2 shows the complete hierarchy of relations specified in the *RBO* module.

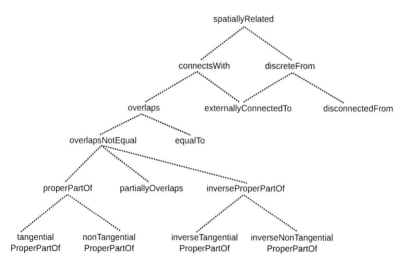

Figure 4.2: *RBO* **Module: Relation hierarchy.** RCC-8-based relations between regions and their hierarchical dependency. Dotted lines indicate subrelations.

As the development of an RCC-8 ontology specification is directly influenced by the RCC-8 spatial calculus itself, no testing and adjustment of the module is required except for an evaluation that the *RBO* module complies with the RCC-8 definitions. Clearly, no composition table is available from the *RBO* module, but it specifies the 8 basic relationships as ObjectProperties, their characteristics, and hierarchical property dependencies. This hierarchy is structured according to the "subsumption lattice of dyadic relations" [Cohn et al., 1997a] of RCC-8 except for the equality relationship equalTo, which is directly subsumed by overlaps according to Grütter et al. [2008]. As analyzed above, the JEPD requirement between the basic relations is in principle satisfied. Furthermore (although not directly relevant for the *RBO* module), as the *RBO* module defines the 8 basic relations of RCC-8, the module implicitly has a cognitive bias, as the granularity of these 8 relations is clearly distinguishable by humans [Renz et al., 2000].

Table 4.1 shows an overview of the technical specification of the region-based ontology module *RBO*. Its applicability is indirectly tested by its use in other spatial modules, which is described in section 6.2, where it has also been used with spatial and DL reasoning.

4.2.1.2 Architectural Construction Ontology

The Architectural Construction Ontology (*ACO*) is a module that specifies constructional building elements as they are used in blueprints for architectural design. The entities that are illustrated on constructional building floor plans, such as walls, doors, pieces of furniture, or stairs, are examples of such elements, that are accordingly specified in the *ACO* module. The entities are primarily defined by their quantitative properties, size and position. In particular, the properties reflect metric data of construction elements in building blueprints, e.g., heights of ceilings, positions of walls, widths and heights of windows, or opening angles of doors.

The purpose and aim of this ontology module is to reflect spatial metric information about constructional elements in architectural plans. The spatial perspective is thus primarily based on quantitative categorizations in the context of architectural designs that take into account the semantic types of the elements in such designs. Hence, the module specifies constructional architectural entities from a quantitative perspective, i.e., metric aspects about the elements are primarily described. The *ACO* module is also intended to be used as an intermediate abstraction layer between the construction's raw

Table 4.1: **Details of the Region-Based Ontology Module** *RBO.* Specification overview of the ontology module that defines the basic relations of RCC-8.

	Region-Based Ontology Module
Ontology Module	$RBO = \langle C_{RBO}, R_{RBO}, In_{RBO}, A_{RBO} \rangle$
Spatial Perspective	SP_{RBO} : Qualitative perspective (region-based)
Concepts	C_{RBO} : No specific categories distinguished except for regions
Relations	R_{RBO} : RCC-8-related spatial basic relations together with their hierarchical dependencies
Instances	In_{RBO} : Regions and their region-based relations
Axioms	A_{RBO} : Property hierarchy and characteristics for RCC-8-region-based ObjectProperties
Purpose and Aim	Ontological specification of RCC-8 to provide region-based relations for other spatial modules
Requirement	Compliance with RCC-8
Example Application	Reused and applied in other spatial modules (for example, section 6.2)
Development	Use of RCC-8 specification [Cohn et al., 1997b] and extension of RCC-8 specification by Grütter et al. [2008]
Ontology Language	DL (OWL 2), supported by DL reasoners
Expressivity	$\mathcal{ALRI}(+)$, 2 classes, 15 relations
Availability	Available online at `http://www.informatik.uni-bremen.de/~joana/ontology/modSpace/RCC-Ontology.owl` (cf. appendix A)

data (represented in some architectural data format, e.g., a CAD model) and abstract conceptualizations of architectural design (represented, for instance, as an instantiation in the *ADO* module, cf. section 4.2.3.2). Hereby, the module grounds the ontological categories into constructional data.[6]

Applications that can use this ontology module are design tools for generating architectural building designs. Nowadays, applied CAAD (computer-aided architectural design) tools still lack the possibility to reflect and reason with spatial semantic categories [Hois et al., 2009b]. The spatial ontology specification required for architectural design applications has to reflect information about the actual floor plan, its relevant elements, and its quantitative spatial information. Building requirements formulate constraints on the floor plan. Whether a specific floor plan satisfies the requirements can then by proven by using ABox reasoning over the floor plan's instantiation in the *ACO* module. For example, an architectural constraint may require that all doors in a floor plan design have a certain minimal size for accessibility reasons. This constraint, specified in the *ACO* module, has to be satisfied by concrete floor plan instantiations of the *ACO* module accordingly. In section 6.2, an assisted architectural design application is presented, which implements even more complex design guideline requirements by reusing the *ACO* module. The scenario enables the identification of architectural design plans that satisfy certain design guidelines by using ontological reasoning.

As the *ACO* module reflects information about constructional architectural plans, it is evident that the module should be related to standards for architectural design to facilitate interoperability with computer-aided architectural design tools [Kalay and Mitchell, 2004]. Following conventions and reusing standardized design formats improves applicability and distribution of the *ACO* module. One of the major architectural design formats is the *Industry Foundation Classes* (IFC), here used in version 2x edition 4 release candidate 2 [Liebich et al., 2010], which is a "data schema used to exchange and share structured building information among various software applications used in the construction and facility management industry sector" [Froese et al., 1999]. IFC can be related to 3D CAD models, though its data model is more expressive than CAD [Liebich et al., 2010]. It defines not only geometric primitives, such as points, lines, and polygons, and raw metric data about these entities, it also defines primitive

---

[6]As architectural building plans are a specific type of maps, the *ACO* specification may be reused or adjusted for other similar map-like representations, which is, however, left for future work.

semantics for them by relating these objects to structural elements: it defines concrete building components such as walls, windows, or roofs, as well as abstract entities such as actions, spaces, or costs. The XML data format of IFC is supported by commercial as well as free software design tools, which also facilitates its applicability for modeling, visualizing, or syntax checking [Hois et al., 2009b]. However, the IFC data format is not available as an ontology or as a DL specification in OWL. The translation from IFC types into an ontological specification together with an ontological structuring of these types is achieved by the *ACO* module.

IFC specifies different types of *building entities* that provide a basis for an ontology module on the level of basic quantitative spatial information. The *ACO* module consequently resembles relevant IFC classes necessary for specifying structural aspects of an architectural design.[7] IFC provides the different architectural entities of a design and their basic properties. For instance, a door in IFC is defined as *IfcDoor* that has (i) a width and height, (ii) an opening direction, and (iii) that is connected with other walls or windows (*IfcRelConnectsPathElements*). This information is accordingly encoded in the *ACO* module, as specified in listing 4.4.

Listing 4.4: Specification of the category Door as a constructional building element in the *ACO* module

Class: Door

    Annotations:
        rdfs:comment ' *Door types defined as constructional building elements.* ' *@en*

    SubClassOf:
        StructuralBuildingElement ,
        openingAngle **exactly** 1 float ,
    rdfs:comment ' *Adjacent constructional elements of door instances have to be at least one of type Wall or Window.* ' *@en*
        connectedWith **some** (Wall **or** Window),
    rdfs:comment ' *(The following axioms are inherited from the super category StructuralBuildingElement, shown for reasons of clarity.)* ' *@en*
        height **exactly** 1 float ,

---

[7]As the types that IFC provides go beyond the purpose of the *ACO* module, e.g., information about procurement and cost estimation, only the constructional and structural IFC types are reused in this module.

```
length exactly 1 float ,
width exactly 1 float
```

Information provided by this quantitative module is related to concrete floor plans and their constructions are instantiated accordingly. The module can also be refined, e.g., by specifying minimal or maximal sizes of certain constructional entities (e.g., rooms or corridors) on a metric basis. For this purpose, the refinement can use concrete domains [Haarslev and Möller, 2003] for specifying the metric ranges of entities (cf. section 6.2). Hence, the values of the size of certain entities, such as windows and doors, can be bound to certain upper and lower limits, e.g., to ensure accessibility.

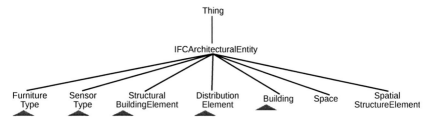

Figure 4.3: *ACO* **Module: Constructional Architectural Entities.** Main category distinction in the *ACO* module, resembling IFC's architectural entities. Triangles indicate further subcategories. FurnitureType, for instance, is subdivided into Clock, Locker, Mirror, Sideboard, etc.

Figure 4.3 illustrates the main categories in the *ACO* module. The category comprising architectural entities is a subclass of the DL's top node Thing and subsumes the following categories that resemble, modify, and refine IFC-related types:

FurnitureType. Subcategories of this type resemble what is identified by IFC's IfcFurnitureType with more structural definitions, i.e., subclass relationships have been introduced that are not specified by IFC. For example, the types Board, Blackboard, DryMarkerBoard, and PinBoard are equivalent (on the same hierarchical level) in IFC, but the last three are subsumed by the first in the *ACO* module. This gives a more fine-grained hierarchical distinction among the categories specified by the module. Instances of FurnitureType are contained in a Building instance. Examples are Cupboard, Mirror, Sculpture, and Wardrobe.

StructuralBuildingElement. Subcategories of this type resemble what is identified by IFC's IfcSharedBuildingElement, which defines the major elements that constitute an architectural design of the building structure. StructuralBuildingElement specify the main components of a raw building and its instances have to define metric information about height, width, and length. In contrast to IFC's Ifc-SharedBuildingElement, StructuralBuildingElement does not subsume enumerations of shared building elements. In the ontology module, these enumerated types are specified as properties ranges of StructuralBuildingElement. For instance, IFC's IfcWallTypeEnum is specified as a DataProperty wallType of Wall, which can have one of the following ranges: {STANDARD, POLYGONAL, SHEAR, ELEMENTEDWALL, PLUMBINGWALL, USERDEFINED, NOTDEFINED}. Specifying this kind of information as property ranges more closely reflects the IfcWallTypeEnum, as they do not introduce new subcategories but characteristics of one chategory. Examples of StructuralBuilding-Element subcategories are Ceiling, Door, Roof, and Window.

SensorType. This category specifies types of sensors that are used for detection devices in buildings, for instance, for building automated control systems. It reflects the sensor types defined by IfcSensorType in IFC. A sensor has to be spatially connected to some structural building element. Examples are FireSensor, LightSensor, MovementSensor, and TemperatureSensor.

DistributionElement. According to IFC's IfcDistributionElement, this category specifies electrical, sanitary, or other element types within a distribution system, that are containedIn a Building. It may define a common material or common shape property by the properties materialType and shapeType. Examples of DistributionElement subcategories are CommunicationElement, CoolingElement, HeatingElement, and Plumbing-Element.

Space. According to IFC's Space, this category represents an area or volume that is bounded to a certain region in the architectural design. Space is an area or volume that reflects a region that provides certain functions within a building. All instances of the category thus have to define a function. The category does not have any further subcategories.

Building. This category reflects the entire building site. It is subdivided into three categories. The first category refers to a ComplexBuilding spanning over several connected or disconnected buildings and specifies a collection of buildings included in the site. The second category refers to an ElementBuilding specifying exactly one building. And the third category refers to a PartialBuilding specifying (vertical) building sections.

SpatialStructureElement. This category resembles what is identified by IFC's IfcSpatialStructureElement. Instances of this type are sets of spatial elements that might be used to define a spatial structure. By this, complex elements of any kind that are composed of structural entities can be specified in a flexible way as a building project may require. Instances of this category are thus composed of some relatedObjects, such as Building, DistributionElement, StructuralBuildingElement, or FurnitureType instances. The category SpatialStructureElement thus acts as an aggregate for other building structural entities and it does not have any subcategories.

Basic axioms about the composition of buildings can thus be defined on the basis of the categories in the ACO module. For example, all instances that reflect a single building necessarily have to be related to some constructional and distributional building elements, namely StructuralBuildingElement and DistributionElement. This is ensured by the specification of the category ElementBuilding. The respective specification is presented in listing 4.5. In this way, more specific building requirements can be specified, and refined for concrete building guidelines (cf. section 6.2).

Listing 4.5: Specification of the category ElementBuilding as a composition of constructional and distributional building elements in the ACO module

```
Class: ElementBuilding

    Annotations:
        rdfs:comment  'Instances refer to exactly one building.' @en

    SubClassOf:
        Building,
        contains some DistributionElement,
        contains some StructuralBuildingElement
```

Although the *ACO* module strictly employs the physical perspective and specifies its categories by their metric properties, it also contains some containment object properties and type data properties (e.g., materialType or wallType) as they are closely related to the IFC data model. These types of information, however, do not provide any conceptual information or functional aspects of *ACO* module categories, which is not a weakness of the module but complies with IFC and the spatial perspective adhered to. In order to achieve a description of conceptual aspects of a building in addition to the structural aspects, the categories in the *ACO* module can be related to other modules with abstract or domain-specific perspectives. For example, the *ACO* module can be related to physical objects as specified in the *PEO* module (see section 4.2.3.1). In particular, the architectural construction ontology can ground *PEO*'s physical entities into quantitative data models.

As the *ACO* module reflects merely quantitative information about types from the architectural construction domain, its containedIn and connectedWith ObjectProperties do not have any qualitative semantics but are based on metric information. A qualitative spatial description of this, for instance, a region-based interpretation, can be achieved by connecting the module to the qualitative region-based *RBO* module (see section 4.2.1.1). The *ACO* module then reuses the *RBO* module and specifies that contains and connectedWith (in the *ACO* module) are equal properties to inverseProperPart and connectsWith (in the *RBO* module) respectively. This interpretation from a metric data model to its region-based specification can be automatically determined [Bhatt et al., 2009].

Closely related to the *ACO* module are domain-specific ontologies about architectural design specifying high-level functional and artifact-based aspects of a design. For example, the actual function that a building has to provide, e.g., office building or factory, can be precisely specified by the types of entities the building has to contain and by further requirements the architectural design has to satisfy. This kind of domain-specific information is, for instance, reflected in the architectural design ontology module, presented in section 4.2.3.2. This module reflects conceptual and functional design guidelines; the *ACO* module reflects the constructional building components.

An overview of the overall architectural construction ontology module *ACO* is given in table 4.2. Its applicability and appropriateness in the context of spatial assistance systems is evaluated in chapter 6. For application reasons, the module is connected with several other spatial modules, presented in section 5.1.

4.2.2 Ontology Modules for Abstract Space

Ontology modules that provide spatial specifications from an abstract space perspective address rather general axiomatizations of space, which are often already included in upper-level ontologies (see section 2.6.3). Spatial applications, however, may need a refined version of these representations that match their specific requirements, as the upper-level specifications are too abstract. Nevertheless, these specifications provide a general categorization that can be reused for application-specific refinements. Some foundational ontological specifications may, however, also be reused directly by application-specific ontology modules if they provide enough structure for some spatial domain. For example, the *PEO* module (introduced in section 4.6) refines the foundational ontology DOLCE (cf. section 2.6.3 for an overview) by specifying categories for physical and environmental entities. In the following, an ontology module is presented that specifies general structures of spatial actions.

4.2.2.1 Spatial Actions Ontology

The Spatial Action Ontology (*SAO*) is a module that specifies spatial actions and entities that are involved in the change of spatial types or properties. Hence, requirements of this module are the specification of spatial actions, actors, effects, and combinations of actions (activities), as far as they are relevant for spatial environments. As this module defines spatial information from an abstract perspective, it is not directly linked to or depending on some application. However, the module is further refined and applied to spatial assistive systems, in particular, it is used by the *HAO* module (section 4.2.3.3). The aim of the *SAO* module thus is to formalize a high-level specification for spatial actions and their properties that can be applied in specific domains.

Actions are commonly categorized as subclasses of events: they are more specific than events because there has to exist an agent that performs the action intentionally [Lowe, 2010]. In contrast to actions that are performed by intention, another type of event is the category phenomenon that reflects a happening without intention. Often, a distinction is made between actions and activities, although this definition is not absolute and others have been used [Sandewall, 2006]: actions are primitive events

Table 4.2: Details of the Architectural Construction Ontology Module *ACO*.
Specification overview of the ontology module that defines constructional entities in architectural designs.

	Architectural Construction Ontology Module
Ontology Module	$ACO = \langle C_{ACO}, R_{ACO}, In_{ACO}, A_{ACO} \rangle$
Spatial Perspective	SP_{ACO} : Quantitative perspective (architectural construction metrics)
Concepts	C_{ACO} : Constructional objects in architectural building plans
Relations	R_{ACO} : Metric relationships among constructional objects
Instances	In_{ACO} : Architectural design data (e.g., represented by the IFC data model)
Axioms	A_{ACO} : Class hierarchy and definitions (class constraints) for constructional architectural entities
Purpose and Aim	Description of entities in architectural design plans related to metric aspects
Requirement	Provide interface between architectural data and ontological specification; adhere to some architectural design data format
Example Application	Architectural design and assisted living (see section 6.2)
Development	Reuse of Industry Foundation Classes 2x edition 4 release candidate 2
Ontology Language	DL (OWL 2), supported by DL reasoners
Expressivity	$\mathcal{SRIQ}(D)$, ~190 classes, ~30 relations
Availability	Available online at `http://www.informatik.uni-bremen.de/~joana/ontology/modSpace/ArchitecturalConstruction.owl` (cf. appendix A)

that cannot be split into further sub-actions; activities are (ordered) sets of actions. Activities may also be used to define plans in order to achieve a certain goal. This distinction between actions and activities can certainly be applied to spatial actions and it is used in the *SAO* module.

Aspects of actions contain the actor that intentionally performs an action, actees that may be affected by the action, instruments that are used to perform the action, manners in which the action is performed, and circumstantial information about the action, e.g., when and where it takes place [Davidson, 1967]. In contrast to general actions, spatial actions, necessarily require spatial aspects in real space. For example, the action 'thinking' may take place in someone's mind, but except for this placement no actual spatial change is involved in this action.

In ontological research, spatial actions such as movement, i.e., an object changes its spatial positions over time, depend accordingly on definitions of time, space, objects (actors and actees), and positions [Galton, 1997]. Respective theories are commonly formalized in first- or higher-order logics, but as application-specific modules presented below (see section 4.2.3.3 for a home automation ontology) require decidability and technical applicability to be used in concrete spatial assistive environments, a DL representation is more convenient. The *SAO* module thus has to provide such a representation that contains the formal specification of spatial actions.

An upper-level ontology that can be reused for this purpose, as it already specifies the general information about actions and activities, is an "Ontology of Descriptions and Situations" (DnS) [Gangemi and Mika, 2003], which is formulated in OWL. DnS provides a framework for representing contexts, methods, norms, theories, and situations. Among many other categories, it specifies activities and planning. The ontology also reuses categories from the foundational ontology DOLCE[8]. The following categories from DnS are particularly relevant for describing spatial actions and are further refined in the *SAO* module.

Action. In DnS, the category action is specified as an event that "exemplifies the intentionality of an agent"[9], and thus fits the above-mentioned definition. All instances

---

[8]In section 4.2.3.1, an introduction to DOLCE's categories is presented, which is is not relevant here for the description of the DnS ontology and the *SAO* module.

[9]http://www.loa-cnr.it/ontologies/ExtendedDnS.owl

of action have to be related to some agent and they have to indicate temporal information (when or for how long the action takes place). The *SAO* module introduces SpatialAction as a new subtype of DnS's action category that additionally requires the spatial action to provide information about the spatial change of objects. In detail, every spatial action has a spatial effect on some other object, as illustrated in listing 4.6[10]. Examples of a spatial action that are refined for applicability in spatial assistive systems, which is described in section 4.2.3.3, are DoorOpening or RoomEntering.

Listing 4.6: Specification of spatial actions in the *SAO* module

```
Class :  SpatialAction

    Annotations :
            rdfs : comment    ' spatial  action  category  ( its  subcategories
            can  be  used  in  assistive  system  environments ). ' @en

    SubClassOf :
            ExtendedDnS : action ,
            actor some  SpatialObject ,
    rdfs : comment    ' the  effect  can  be  on  any  object  in  the  environment
    that  is  defined  by  a  physical  object ,  which  could  be  the  actor ,  actee ,
    or  circumstantial  objects  of  the  action . ' @en
            spatialEffectOn some  DOLCE–Lite : physical –object ,
```

Activity. According to DnS, an activity is a set of actions that are part of this activity. This can be directly applied for spatial activities. Hence, the category SpatialActivity of the *SAO* module is introduced as a subclass of activity of the DnS ontology that can only consist of SpatialActions. It consequently has to define SpatialActions as parts of the activity. The actions of an activity can also be ordered by using the relation sequenced-by of the activity category.

Physical Object. As shown in listing 4.6, spatial objects perform the action as an actor. These objects are introduced in the *SAO* module, in order to refer explicitly to the intentional agent. This category is specified by SpatialObject, which is a subcategory of physical-object. This way, a spatial assistive application can individually refine this

---

[10]Although the namespace and the URI of the DnS ontology is 'ExtendedDnS', it refers to the current specification of DnS with an extended vocabulary for social reification.

category and apply it to its domain-specific environment, e.g., particular user groups can be further defined that are allowed or able to perform certain actions.

An overview of the spatial action ontology module *SAO* is given in table 4.3. Its applicability is indirectly tested by the refinement for spatial assistive systems, presented and discussed in section 6.2.2, where the *SAO* module is primarily used for representing user-specific actions that can be performed in indoor apartment environments.

4.2.3 Ontology Modules for Domain-Specific Space

Ontology modules that specify spatial information from a domain-specific perspective are closely related to the requirements of spatial applications. They are designed to reflect spatial information in the application's domain, mostly categorizations of environmental aspects. These modules may reuse ontology modules from the abstract space perspective and further refine them with regard to their use and adequacy in spatial systems. In the following, three domain-specific spatial ontology modules are presented that specify information about spatial physical entities, conceptual architectural design, and assistive home automation respectively.

4.2.3.1 Physical Entities Ontology

The Physical Entities Ontology (*PEO*) is a module that specifies physical, concrete, real-world entities, which occur in space. These entities can also be understood as those objects that are visually perceivable. Thus they can technically be grounded in environmental data, i.e., instances of the ontological categories can be linked to (grounded in) data that reflect their related real-world objects. The main purpose of the *PEO* module is to support spatial systems to perceive, recognize, and classify their environments and the objects they contain. The ontological specification aims at improving the visual classification that is based on low-level image processing. This processing is often ambiguous, in which case the ontological specification adds further information about environmental interrelations that can support the classification process. Thus, the module has to provide a description of the objects, their properties, and their spatial and functional relations in the environment. This description also has to take into account perceptual and ideally cognitively appropriate features of the objects and their relations.

Table 4.3: **Details of the Spatial Action Ontology Module SAO.** Specification overview of the ontology module that defines spatial actions from an abstract space perspective.

	Spatial Action Ontology Module
Ontology Module	$SAO = \langle C_{SAO}, R_{SAO}, In_{SAO}, A_{SAO} \rangle$
Spatial Perspective	SP_{SAO} : Abstract space with respect to spatial change and actions
Concepts	C_{SAO} : Spatial actions and their participants and effects
Relations	R_{SAO} : Properties of spatial actions
Instances	In_{SAO} : Domain-specific refinements for spatial applications that introduce concrete spatial events in spatial assistive systems and that can be instantiated according to the environmental change
Axioms	A_{SAO} : Action-related constraints
Purpose and Aim	General specification of categories that are involved in spatial actions and that can be refined for specific application purposes
Requirement	Provide a general formalization for categories about spatial actions that can be applied within spatial systems
Example Application	Refinements of the SAO module are used for spatial assistive living environments (see section 6.2.2)
Development	Reuse of foundational ontology DnS; ontology categories and relations are added for specifying spatial actions
Ontology Language	DL (OWL 2), supported by DL reasoners
Expressivity	$\mathcal{SHIN}$, ~100 categories and ~230 relations (high number due to DnS import)
Availability	Available online at http://www.informatik.uni-bremen.de/~joana/ontology/modSpace/SpatialAction.owl (cf. appendix A)

The specification of the *PEO* module is guided accordingly by results from empirical research and spatial cognition, in order to reflect perceptual aspects for supporting the visual classification. For example, the module makes a basic distinction between indoor and outdoor environments, as they are perceived differently by humans [Klippel et al., 2006]. Consequently, indoor environments differ from outdoor environments in their ontological axiomatization: The indoor environments (e.g., kitchen, office, laboratory) are specified by explicit constraints about their internal structure and the types of objects they contain, the outdoor situations (mountain scenery, pavement, parking lot) have less clear boundaries and are less constrained with respect to their visual structure and possible objects they contain [Reineking et al., 2009].

An application that can use the *PEO* module is, for instance, a robotic agent that navigates through space and that has to recognize its environment. Such an agent can explore its surroundings by using sensorimotor abilities to detect environmental information. Here, domain-specific knowledge that is specified employing the *PEO* module cannot only be used for describing the objects the robot comes across but also the consequences that can be inferred from the (spatial) configuration of these perceived objects. If, for instance, the robotic agent detects a certain number of objects that are all related to home furniture types and electrical devices, it may conclude that it is located in a domestic environment, which may affect the robot's planning, recognition, or further actions. The *PEO* module has been specifically developed for vision recognition systems of such spatial agents that use the ontology for visual object and scene classification, which is described in section 6.3.

On a coarse level, the *PEO* module specifies the categories building types, such as universities and office buildings (UniversityBuilding, OfficeBuilding), outdoor types, such as sidewalks, streets, and mountains (Sidewalk, Street, Mountain), and physical (visually perceivable) objects, such as types of furniture, traffic lights, and plants (FurnitureEntity, TrafficLight, Plant). Constraints about possible combinations of these objects in spatial environments and resulting consequences are determined by particular spatial applications, for instance, for the indoor environment scenario described in section 6.3. Here, a Kitchen is defined by containing (contain) at least one of the object types Oven, Sink, or Refrigerator, besides other appliances.

As the *PEO* module specifies environmental objects that describe concrete, real-world entities, it is based on (i.e., it reuses) existing foundational or upper-level ontologies that provide a general structure of the physical domain. Ideally this ontology should further provide a perception-based or cognitively influenced description of this domain, as the *PEO* module is intended to support the vision recognition process of spatial agents. An upper-level ontology that specifies this kind of general domain structure is the DOLCE[11] ontology [Masolo et al., 2003]. The benefits of reusing DOLCE are its specification flexibility, cognitive adequacy, and development guidance, as discussed in the following overview of DOLCE.

DOLCE is a "Descriptive Ontology for Linguistic and Cognitive Engineering", which aims at providing an "explicit representation of the so-called ontological commitments about the meaning of terms, in order to remove terminological and conceptual ambiguities" [Masolo et al., 2003, p. 2]. The categories specified in DOLCE have been strongly influenced by a cognitive bias in the development process, i.e., the categories reflect information about cognitive artifacts depending on human perception, cultural imprints, or social conventions. DOLCE can be used to make rationales explicit, i.e., its structure can guide the development of new ontologies. As a consequence, the ontology can be used as a reference point for comparison of different ontologies or as a mapping aid between these ontologies. This mapping aid has also been applied for combining the *PEO*, *SAO*, and *HAO* modules in an assisted living system, presented in section 6.2. With regard to its applicability, DOLCE is also designed to be reused for domain-specific ontologies that specify concrete physical real-world entities. It is thus particularly relevant to the *PEO* module, and this is the main reason why the module reuses DOLCE.

In DOLCE, categories are specified in a *minimal* way [Masolo et al., 2003, p. 94], i.e., it defines category types as general as possible. This enables fine-grained specifications in the *PEO* module about physical spatial environment, and the distinctions between indoor and outdoor entities or scenes and objects to be introduced by refining the DOLCE taxonomy (see below). As DOLCE provides category specification not only in a most general but also in a *rigorous* way [Masolo et al., 2003, p. 4], categories are richly axiomatized. As a consequence, the *PEO* module can build on this axiomatization by reusing most of DOLCE's relations. Only few subtypes of ObjectProperties in DOLCE have to be newly specified in the *PEO* module. Although DOLCE describes only categories that are *particulars* (see section 4.1.1), i.e., its categories cannot have

---

[11]http://www.loacnr.it/DOLCE.html

instances, the *PEO* module specifies universal categories that can be instantiated. For instance, DOLCE's category physical-object (see below) cannot be instantiated, but subtypes of it introduced in the *PEO* module can, such as sofa, book, plant, or traffic light.

DOLCE's generalized specification together with its detailed axiomatization supply a high flexibility for the creation of new ontologies that refine DOLCE's categories, i.e., new ontologies can individually introduce new subcategories guided by the axiomatization of DOLCE's supercategories. Its thorough axiomatization can thus improve the specification of refinements of categories and relations in the *PEO* module. Moreover, as DOLCE's categories are influenced by human perception and social conventions, it is particularly appropriate for specifying real-world entities in a physical domain. DOLCE is thus reused for the physical entities ontology *PEO*. To support computational applicability, the *PEO* module imports the DL version of DOLCE, namely DOLCE-Lite, which is available in OWL [Gangemi and Guarino, 2004].

On the top-most level, DOLCE-Lite (as well as DOLCE) makes the distinction of the four types Endurant (Continuants), Perdurant (Occurrents), Quality, and Abstract. Endurants are continuous entities that exist in time. These are primarily physical entities with physical, i.e., spatial, extensions. Examples of endurants are a hammer, some gold, or a house. Perdurants are occurring entities that happen in time. These are primarily events and processes. Examples of perdurants are death, conference, sitting, or running. Qualities are basic entities that can be perceived or measured. They inhere in other endurants, perdurants, or qualities. Examples are shapes, colors, sizes, or sounds. Abstracts are entities that define values or a value space for qualities. These values can change over time, but the quality itself remains fixed. Examples of abstracts are the redness of a rose's color, or the round shape of a shape quality. Figure 4.4 shows the category hierarchy of the DOLCE taxonomy. The *PEO* module refines subcategories of endurants and qualities (see below).

Some of the main DOLCE relations that are reused by the *PEO* module are parthood and temporary parthood, dependency and spatial dependency, constitution, and participation, quality inherence and quality value. An example of the axiomatization of the category physical-endurant in DOLCE-Lite is shown in listing 4.7. The specified category is a subcategory of endurant that can only have physical qualities, that has to specify a

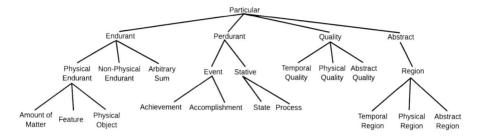

Figure 4.4: DOLCE Taxonomy. The most general categories in the hierarchy of the DOLCE taxonomy (an excerpt taken from the complete taxonomy presented by Masolo et al. [2003]).

spatial location, that can have parts only of the type physical-endurant, and that can only constitute (be part of) other physical endurants.

Listing 4.7: Specification of physical endurants in DOLCE-Lite (comments in the subclass description have been added for intelligibility)

```
Class :  physical −endurant

    Annotations :
          rdfs : comment   ' An  endurant  having  a  direct  physical  (at  least
          spatial )  quality . ' @en

    SubClassOf :
          endurant ,
    rdfs : comment   ' A  physical  endurant  can  only  define  physical
    qualities .  More  specifically ,  it  has  to  define  at  least  one
    spatial  location  (in  an  ordinary  space ,  e.g. ,  coordinates ). ' @en
          has−quality  only  physical −quality ,
          has−quality  some  spatial −location_q ,
    rdfs : comment   ' (The  following  requirement  is  already  satisifed
    by  the  previous  line ,  still  defined  in  DOLCE−Lite .) ' @en
          has−quality  some  physical −quality ,
    rdfs : comment   ' A  physical  endurant  can  only  have  parts  that
    are  also  physical  endurants . ' @en
          part  only  physical −endurant ,
    rdfs : comment   ' If  a  physical  endurant  is  a  constituent  of  a  setting
    ( situation ) ,  this  setting  can  only  be  another  physical  endurant
    (e.g. ,  some  amount  of  clay  is  a  constituent  of  a  vase ). ' @en
          specific −constant−constituent  only  physical −endurant
```

DisjointWith :
rdfs : comment   ' A physical endurant is disjoint with the other
subcategories of endurant, namely non−physical endurants
(entities with no mass) and arbitrary−sums (collections of
arbitrary endurants). ' @en
    non−physical−endurant ,
    arbitrary−sum

The physical entities ontology *PEO* reuses and refines the categories non-physical-endurant, physical-object, and physical-quality (illustrated in figure 4.5), and the relations generic-location and generic-location-of of the DOLCE-Lite ontology, as they are the relevant types of entities and relations that occur in the physical entities domain. As the visually perceivable real-world entities exist in time, they are subsumed by DOLCE's physical endurant. Their properties are specified by subtypes of DOLCE's quality categories. Furthermore, as collections of the real-world entities can produce socially-constructed new types of entities (e.g., kitchens or offices), these are subsumed by non-physical endurants that depend on certain physical endurants. As a result, the categories physical-object, non-physical-endurant, and quality are further refined and distinguished in the *PEO* module, which is described in the following sections.

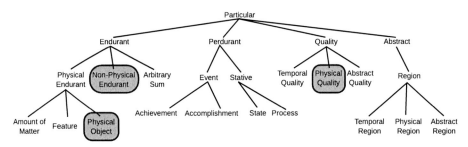

Figure 4.5: Refined DOLCE Categories. Categories marked with boxes in the DOLCE taxonomy are refined by the *PEO* module.

Physical Object. In DOLCE, physical objects are endurants that satisfy the unity criterion, i.e., if an object can be split up into parts, all of these parts are members of the object and they are only connected to each other but not to any other parts [Guarino and Welty, 2002]. Some of these members may change over time, but the object itself keeps its identity. Hence, physical objects can have temporary parts. Physical objects also have direct spatial properties, in particular, they have a location in space and a spatial extent. They are independent from occurrences, i.e., no specific events are required to allow a physical object to exist. This definition then comprises the indoor and outdoor entities existing in spatial environments, such as Chair, TrafficLight, Ceiling, or Pedestrian.

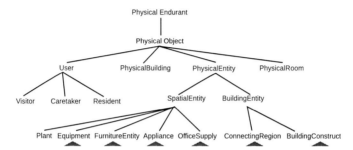

Figure 4.6: *PEO* Module: Physical Objects. Subcategories of DOLCE's physical object category, specified as a refinement in the *PEO* ontology module. Triangles indicate further subcategories. FurnitureEntity, for instance, is subdivided into Desk, Shelf, Sofa, etc.

In the *PEO* module, the physical object category of DOLCE is further subdivided into User, PhysicalBuilding, PhysicalEntity, and PhysicalRoom. This sub-hierarchy is shown in figure 4.6. These types of categories occur in indoor environments, and the specification can thus be used for spatial navigating of systems through these environments (see section 6.3). User captures information about possible user groups a spatial agent may come across in an assisted living environment. This information is relevant for automated systems and applied in the assisted living application, described in section 6.2. PhysicalBuilding reflects the concrete building structure, i.e., the physical and material parts of the building. This category is independent of the actual type of the building, e.g., office building or factory, and thus refers only to the structural building entity. The way this building is used (i.e., as an office building or factory), is defined by the category BuildingType, as described below. A PhysicalBuilding can contain PhysicalRoom entities, that are defined by their physical properties, namely their physical extent and boundaries. An instance of PhysicalRoom can only be contained in one PhysicalBuilding.

This also implies that a PhysicalBuilding may not contain any instance of a PhysicalRoom, which is, for instance, the case for hangars. This class definition uses the containedIn relation, which is introduced in the *PEO* module as a transitive subtype of DOLCE's generic-location. The class definition for PhysicalRoom is formulated by the axiomatization shown in listing 4.8.

Listing 4.8: Axiomatization of the PhysicalRoom category in the *PEO* module that allows a physical room to be part of only one physical building

Class: PhysicalRoom

> **Annotations:**
>> *rdfs:comment* ' *Physical rooms are defined by some physical boundaries, e.g. walls. They can play the role of a room function (kitchen, lab, etc.). Physical rooms contain physical entities. Instances of room types depend on instances of physical rooms.* ' @en

> **SubClassOf:**
>> dolceLite:physical−object ,
>> *rdfs:comment* ' *The relation containedIn is specified in the PEO module as a subrelation of DOLCE−Lite:generic−location , which describes a generalized position of any particular. The position is defined by another endurant, namely a PhysicalBuilding (in contrast to the quality spatial−location_q , which is defined for ordinary spaces).* ' @en
>> containedIn **exactly** 1 PhysicalBuilding

> **DisjointWith:**
>> User ,
>> PhysicalEntity ,
>> PhysicalBuilding

The category PhysicalEntity in the *PEO* module is subdivided into the categories BuildingEntity and SpatialEntity. BuildingEntity specifies non-detachable constructional and structural parts of a PhysicalBuilding; SpatialEntity specifies detachable interior entities, which are contained in a PhysicalRoom and hence in a PhysicalBuilding. The categories distinction is thus based on characteristics of the objects, whether they are easy to remove (detachable) from the environment (the building or room) or not. Consequently, the former category specifies subcategories related to structural building elements, such as Ceiling, Wall, Doorway, or Staircase, and the latter category specifies subcategories related to interior building elements, such as Lamp, Chair, Pen, or Plant. Although the set of possible types of such objects is most likely infinite, they can, however, be specified according to specific spatial applications: the domain-specific environment determines which types of such entities need to be defined. The scene recognition system, in which such types are specified in an application-driven way, is presented in section 6.3.

User is another subcategory of DOLCE-Lite's physical-object in the *PEO* module. It specifies entities that are (detachable) agentive persons (parts) in a physical building. They can use or interact with physical entities in a building. The *PEO* module defines subtypes of User, such as Caretaker, Resident, and Visitor. The group of such subtypes is again infinite and depends on concrete environments, for instance, assisted living environments as presented in section 6.2.

Non-Physical Endurant. In DOLCE, non-physical endurants exist in time, but they have no mass. They are thus dependent on other objects or agents. In the case of spatial environments, non-physical endurants primarily occur as functional assignments to buildings, parts of buildings, or parts of outdoor sceneries. In DOLCE, these functional types *generically constantly depend* on other physical objects, i.e., at any time the functional type exists the physical objects it depends on exist as well. Functional types are thus limited to the physical boundaries of these objects.

In particular, the *PEO* module specifies specific types of buildings and rooms, illustrated in figure 4.7. RoomType and BuildingType are subdivided into particular functional types of rooms and buildings, e.g., ArtGallery, Museum, and Theater are subtypes of the category CulturalBuilding, which is a subcategory of BuildingType. BuildingType requires to be related by the relation dolce:generically-dependent-on to some PhysicalBuilding; RoomType requires to be related to some PhysicalRoom. Furthermore, instances of RoomType are related to some BuildingType by DOLCE's relation of abstract-location, i.e., the building type provides a placement for the room type. The class definition of RoomType is shown in listing 4.9.

Listing 4.9: Axiomatization of the RoomType category in the *PEO* module that determines the functional characteristics of a room. Instances of RoomType are located in a functional type of a building and depend on a physical room.

Class : RoomType

Annotations :

111

> rdfs:comment 'Room types describe functions of physical
> rooms, e.g. kitchen, lab, etc. Instances of room types
> depend on physical rooms.' @en

SubClassOf:
   dolceLite:non−physical−endurant,
   dolceLite:abstract−location **only** BuildingType,
   dolceLite:generically−dependent−on **only** PhysicalRoom

In the *PEO* module, the specification of functionalities of buildings and rooms is axiomatized as part of the category specifications for RoomType and BuildingType. As a consequence, not every physical building or room necessarily has a relation to a building or room type. This ensures the existence of physical buildings and rooms, for which no functionality exist, e.g., in case of empty and unoccupied rooms or empty houses.

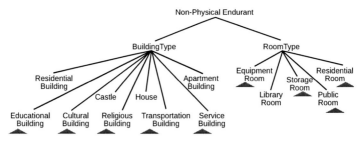

Figure 4.7: *PEO* **Module: Non-Physical Endurant Refinement.** Subcategories of DOLCE's non-physical endurant category, specified as a refinement in the *PEO* ontology module. Triangles indicate further subcategories. TransportationBuilding, for instance, is subdivided into Airport, Harbor, and Station.

Physical Quality. In DOLCE, the category Quality defines values or properties for endurants and perdurants. All endurants or perdurants have qualities that inhere in them as long as they exist. The range of the quality is specified by a certain value space, which can be metric, qualitative, or conceptual. One of the subtypes of Quality is Physical-Quality. This, for instance, defines for each physical object (consequently, for all physical buildings, physical rooms, etc.) its spatial location, e.g., specified by coordinates. The category Physical-Quality is further refined in the *PEO* module. One of such refined subcategories in the *PEO* module is the abstract quality Material. It defines the type of material a spatial entity consists of. A door, for instance, may consist of wooden, metal, or glass material. The set of possible subcategories of Physical-Quality, such as temperature, color, or shape, is again infinite and limited to the application scenarios introduced in sections 6.2 and 6.3.

An overview of the overall physical entities ontology module *PEO* is given in table 4.4. Its applicability and appropriateness in the context of spatial assistance systems is evaluated in chapter 6. Here, the *PEO* module specification for indoor and outdoor spatial environments is used by visual object and scene recognition systems and assisted living environments. Some categories in the *PEO* module, however, may need to be further refined to satisfy new application requirements, which is described in section 6.3.2.

4.2.3.2 Architectural Design Ontology

The Architectural Design Ontology (*ADO*) is a module that specifies spatial information about building structures on the basis of their functional features in architectural designs. In contrast to the *ACO* module, introduced in section 4.2.1.2, it does not specify architectural entities on the basis of their metric properties. Instead, the module specifies entities on the basis of their structural (not constructional) and functional properties. In particular, it aims at reflecting *artifactual* spatial aspects, such as the interaction space of a door, the navigational properties of a floor, or the monitoring aspects of a movement sensor, by defining the related entities and their abstract or qualitative spatial relationships and properties. It furthermore resembles the conceptual specification of an architectural design about element, building, or room types, e.g., living room, storage room, or balcony, resembling the building and room types in the *PEO* module.

The purpose of the *ADO* module is primarily to provide categories that are relevant for conceptual and functional elements in architectural designs. The *ACO* module reflects how a floor plan is geometrically structured; the *ADO* module reflects, what its function is, how the elements may interact, and what new categories are introduced by using certain constructional elements in a floor plan. For example, in the metric representation a door is defined by its size, type, material, and opening angle, the functional representation defines a door as having a functional space, in which another element has to be located in order to interact with the door in a certain way, e.g., to open or shut the door. Although the *ADO* module introduces this functional space as an abstract

Table 4.4: Details of the Physical Entities Ontology Module *PEO*. Specification overview of the ontology module that defines physical entities in spatial indoor and outdoor environments.

	Physical Entities Ontology Module
Ontology Module	$PEO = \langle C_{PEO}, R_{PEO}, In_{PEO}, A_{PEO} \rangle$
Spatial Perspective	SP_{PEO} : Domain-specific perspective (physical indoor and outdoor environments)
Concepts	C_{PEO} : Physical (real-world) objects
Relations	R_{PEO} : Physical relations of concrete objects
Instances	In_{PEO} : Environmental data
Axioms	A_{PEO} : Domain-specific constraints
Purpose and Aim	Description of objects in real world spatial environments; environmental recognition in visual systems
Requirement	Provide knowledge base for spatial application; specify indoor and outdoor objects, their properties and relations; grounded in environmental data
Example Application	vision recognition (see section 6.3), assisted living (see section 6.2)
Development	Reuse of foundational ontology for physical objects (DOLCE-Lite); ontology categories and relations added according to the requirements by application environments
Ontology Language	DL (OWL 2) supported by DL reasoners
Expressivity	$\mathcal{SHIQ}$, ~160 categories and ~70 relations
Availability	Online available at `http://www.informatik.uni-bremen.de/~joana/ontology/modSpace/PhysicalEntity.owl` (cf. appendix A)

artifactual element related to another entity, its actual metric information in the *ACO* module is not directly available, because the space depends on the function it supports and potentially an interactor who uses this function or other context-dependent information. Thus, the functional space is not determined by the metric *ACO* module (and it is not represented in floor plan maps), but it exists on a conceptual level in the architectural design [Lawson, 2001] and is reflected in the *ADO* module by such space categories (see below). However, in a concrete scenario the metric data can be calculated [Bhatt et al., 2009].

Hence, requirements for the *ADO* module are to closely reflect information from the *ACO* module whenever necessary in order to simplify information exchange across these modules, but at the same time formalizing independently its intended specification of the functional building structures and artifactual elements. An application example that is supported by the specification of the *ACO* module is an ambient assistive living environment. Here, the functional and structural features of the environment are of high importance to the designated tasks of assistive living. If, for instance, the environment is supposed to monitor and regulate the temperature automatically (potentially on a user-defined basis), relevant information about the position of the temperature sensing devices, their functional ranges and spatial surroundings, and further properties or objects they may interact with are required for a fully represented situation, on which basis the assistive living environment can then make decisions for temperature regulations. Such an assistive living application scenario is presented in section 6.2.

Figure 4.8 illustrates the main distinction of the categories in the *ADO* module. Building structures and functional structures are distinguished according to their occurrences in architectural designs:

Building Structure. Subcategories of BuildingStructure specify entities that reflect conceptual information about structural aspects in architectural designs. They define buildings and their structural components as well as elements that occur in buildings. The categories formalize parthood relations among rooms, floors, and buildings, as well as building parts, buildings, and building complexes. All subcategories can specify a FunctionalStructure that indicates their potential functional features. Examples of subcategories of BuildingStructure are Corridor, Entrance, Window, and SensoringDevice. One of the sensing devices, for instance, is the category TemperatureSensor, which defines a

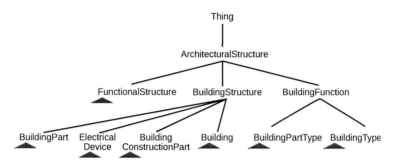

Figure 4.8: *ADO* **Module: Structural Architectural Entities.** Main category distinction in the *ADO* module for structural and functional entities in architectural building environments. Triangles indicate further subcategories. FunctionalStructure, for instance, is subdivided into FunctionalSpace and RangeSpace.

sensor value and a range space, which indicates the functional feature. The specification is shown in listing 4.10.

Listing 4.10: Axiomatization of the TemperatureSensor category in the *ADO* module that determines the functional features of a temperature sensor as a sensing device. Instances of TemperatureSensor specify a range space and some sensor values.

```
Class:  TemperatureSensor

    Annotations:
        rdfs:comment   'A sensoring device that measures the temperature
        inside buildings.' @en

    SubClassOf:
        SensoringDevice,
        rdfs:comment   '(The following axioms are inherited from the
        super category SensoringDevice, shown for reasons of clarity.)' @en
        sensorValue some int,
        hasFunctionalStructure exactly 1 RangeSpace
```

One of the subcategories of BuildingStructure is BuildingConstructionPart, which reflects constructional elements of architectural designs yet with a conceptual focus. For example, such construction-related elements are specified by their metric sizes in the *ACO* module, but the *ADO* module specifies them as having a functional space for interaction or use.

Functional Structure. The category FunctionalStructure specifies entities that provide abstract functional features for other building structure entities, such as the interaction space of a door, the navigational properties of a floor, or the monitoring aspects of a movement sensor. All instances of it have to specify exactly one isFunctionalTypeOf of a BuildingStructure instance. This relation is axiomatized as functional, i.e., for each functional structure there can only exist one building structure element, of which it provides the functional feature. FunctionalStructure is subdivided into RangeSpace and FunctionalSpace. The first specifies ranges for sensoring or other electronic devices; the second specifies functions for building structure types. It can specify the spatial area of a window, in which the user needs to be located in order to interact with the window in a certain way (e.g., opening or closing). It can also specify the area around a lamp which becomes brighter if the lamp is turned on.

On a conceptual level, the functional structure indicates an architectural artifact that reflects a functional feature of an element in an architectural design. The actual size and metric extent of such artifacts can only be inferred in the context of a concrete architectural design. As a consequence, such categories are not reflected in the *ACO* module, but they are specified on a conceptual level in the *ADO* module.

Building Function. Building functions provide functional types of buildings or parts of buildings. The basic distinction into BuildingType and BuildingPartType resembles the distinction of BuildingType and RoomType in the *PEO* module (see figure 4.7). BuildingType specifies functional types of buildings, such as School, OfficeBuilding, or Castle; BuildingPartType specifies functional types of parts of buildings, such as Gallery, StorageRoom, or Library.[12] The *PEO* module defines these types according to their perceptual features and cognitive appearances; the *ADO* module defines these types according to their architectural design functions. In particular, the ontological specification formalizes the category Building and BuildingPart (subcategories of BuildingStructure) to have a relation functionalType to the categories BuildingType and BuildingPartType respectively.

Although the *ADO* module specifies architectural design entities from a spatially domain-specific perspective with an emphasis on functional features of the entities and

---

[12]Building types are taken from the Institute for International Architectural Documentation `http://www.detail.de/rw_3_Archive_De_Gebaeude.htm` (visited on August 1, 2010)

the *ACO* module specifies constructional entities in architectural designs from a quantitative spatial perspective with an emphasis on metric features of the entities, there exist a close relationship between both modules as they both describe architecture-related entities. In architectural design, specific building requirements require constraints on entities from both perspectives, potentially by connecting aspects from both. For example, a concrete design can require that the reception in the entrance area of a public building should have a sufficiently big enough accessibility area around it. In the *ACO* module, constructional information about the entrance area of the building and the reception is instantiated. In the *ADO* module, the functional accessibility area of the reception is instantiated. By connecting both modules and deriving the size of the accessibility area in the *ACO* module, satisfiability of the requirement can be proven. More details about this type of connection is presented in section 5.1. More details about the application in architectural design is presented in section 6.2.

Furthermore, there exists a relation between the *ADO* module and the *PEO* module. Both specify functional types of buildings and their parts. The *ADO* module distinguishes functionalities of building complexes, single buildings, parts of buildings (e.g., levels, wings, sets of rooms, corridors, or entrance areas). The *PEO* module also distinguishes functionalities of buildings and rooms based, however, on their perception-based appearance. Hence, the *ADO* module takes into account architecturally defined types of buildings and building parts and the *PEO* module takes into account agent-based distinctions of visually perceivable functionalities. Furthermore, the *PEO* module formalizes these categories (building and room types) in the same way as it formalizes functionalities of outdoor sceneries, e.g., a road junction or a pedestrian path.

Given the close relationship to the *ACO* module and map-like architectural representations, there is also a link between the *ADO* module and qualitative spatial modules. Functional areas, such as range spaces, for instance, can be reflected as spatial areas around their related structural entities, which can also be reflected by spatial areas. Consequently, their functional space areas and structural entity areas can be formalized as regions with region-based qualitative relationships among them. As discussed in section 4.2.1.1, such representations can be formalized by using the *RBO* module. In a similar way as the relationship between the *ACO* and *ADO* module, the *RBO* module is required to formalize certain architectural design constraints, described in section 6.2.

An overview of the *ADO* module in table 4.5 shows details of its specification. Its applicability and appropriateness in the context of spatial assistance systems is evaluated in section 6.2. Results show that architectural design requirements often have to specify constraints with regard to connections between the *ADO*, *ACO*, and *RBO* modules, and that possible extensions are required for non-standardized architectural designs.

4.2.3.3 Home Automation Ontology

The Home Automation Ontology (*HAO*) is a module that formulates situations or states that are valid for a specific home environment intended to be used in smart homes. It primarily specifies constraints between categories in the environment in the form of property requirements they have to satisfy. The module's aim is to support an automation system for assisted living environments with its ontological specification. It therefore provides information about property ranges and constraints for categories, which are instantiated by the automation system, and it reflects objects in an assisted living environment. If the instantiation complies with the ontological definitions and satisfies all its requirements, the environment's state or condition is valid according to the specification.

Application examples that can make use of the *HAO* module are monitoring systems that control assisted living environments, which supervise environmental conditions and user behaviors and adjust or change states of objects in the environment. If the system, for instance, needs to provide automatic lighting, the monitoring system can use ontological definitions about lighting-related categories and constraints. The *HAO* module is able to define valid lighting conditions, such as "lights should be turned on with a certain intensity in specific places of the environment when users are located in or near these areas". This requirement can be specified by refining class definitions for lights that are located at certain places and by constraining the range of the light intensity property if users are present. The monitoring system can then reason whether its instantiation of the environment satisfies these conditions. Other application examples may contain constraints over activities, emergencies, or security-related situations [Normann et al., 2009].

As a consequence, the *HAO* module is strongly connected to the assisted living environment at hand. A concrete environment is shown in section 6.2, in which the

Table 4.5: **Details of the Architectural Design Ontology Module *ADO*.** Specification overview of the ontology module that defines functional features in architectural designs.

	Architectural Design Ontology Module
Ontology Module	$ADO = \langle C_{ADO}, R_{ADO}, In_{ADO}, A_{ADO} \rangle$
Spatial Perspective	SP_{ADO} : Domain-specific perspective (functional architectural design features)
Concepts	C_{ADO} : Functional and conceptual categories in architectural designs
Relations	R_{ADO} : Spatial-functional relations in architectural designs
Instances	In_{ADO} : Conceptual elements and their functional features in architectural designs plans
Axioms	A_{ADO} : Design constraints
Purpose and Aim	Description of functional features in design plans; task support for conceptualizations in architectural design
Requirement	Provide specification for functional building structures and their artifactual elements
Example Application	Assisted living environments and architectural design (see section 6.2)
Development	Ontology categories and relations taken from architectural design plans
Ontology Language	DL (OWL 2) supported by DL reasoners
Expressivity	$\mathcal{ALCRIQ}(D)$, ~90 categories and ~20 relations, etc.
Availability	Online available at `http://www.informatik.uni-bremen.de/~joana/ontology/modSpace/BuildingStructure.owl` (cf. appendix A)

HAO module is applied for a specific use case. The module has to take into account objects and categories of the environment and their possible conditions together with information about the domain and its instances. For this purpose, the *HAO* module is not only closely related to certain applications that need to define their requirements and valid behaviors, but it also relies on other spatial modules, namely the *ADO*, *PEO*, and *SAO* modules, by reusing their definitions for entities in indoor home environments and their spatial aspects. For this purpose, the *HAO* module is specified in OWL as well.

As requirements may vary across assisted living environments or they may change over time, the module depends on the specific building it is applied to and it needs to be re-defined for different environments accordingly, however, its basic ontological constructions can be reused. The following presentation of the *HAO* module shows aspects that are relevant for the use case described in section 6.2.2, i.e., for an assisted living apartment that provides building automation [Hois, 2010c].

For this application, the *HAO* module defines ontological restrictions on entities from architectural structures from the *ADO* module, physical entities from the *PEO* module, and spatial actions from the *SAO* module by importing and connecting these three ontologies. In general, valid situations in the apartment are specified by refining class definitions and conditional dependencies for entities from all three modules. For example, application-specific constraints over a user (specified in the *PEO* module) who performs an action (specified in the *SAO* module) that affects entities in the environment (specified in the *ADO* or *PEO* module) are formalized in the *HAO* module.

The *HAO* module therefore specifically needs to incorporate several spatial perspectives about its contained objects. For example, the category Door acts as an architectural structure but also as an access control unit. Based on user profiles defined by the assisted living system, the *HAO* module defines who (physical entities from the *PEO* module) is allowed to enter (spatial actions from the *SAO* module) which building element (structural features from the *ADO* module). The technical aspects of integrating these different perspectives by the combination of the ontology modules are explained in detail in section 5.1. The following use case shows the way the *HAO* module formulates environmental requirements for valid situations according to temperature regulations, in which temperature sensors (physical entities) regulate (spatial actions) heaters (physical entities) and enable heating control (home automation).

The assisted living apartment provides the identification of abnormal (invalid) heat or cold. The average temperature range is defined a priori by the system to be within the range of 18–22°C for rooms regularly used by the user, such as living rooms, bedrooms, and bathrooms, and other values for less often used rooms, such as storage rooms. If a temperature value outside this range is detected (the instantiation does not satisfy, e.g., the bathroom categorization), heating should be turned on or off. However, if certain actions take place that strongly affect the temperature, heating control should react differently or (temporarily) ignore the invalid temperature values. For example, when the user is bathing or taking a shower and temperature values above normal are detected in the respective bathroom, the system may ignore this instead of turning on the air conditioning.

Regardless of the particular heating control, the aspects for valid temperatures are specified by the *HAO* module for the assisted living environment. Listing 4.11 shows the specification for bathrooms that contains their valid temperature conditions: Every bathroom instantiation has to have a temperature sensor and it has to comply with the pre-defined temperature values unless a bathing action takes place. If these requirements hold, the instantiation satisfies the specification for being a (environmentally valid) bathroom. This specification in the *HAO* module reuses and extends the definition of bathrooms given by the *PEO* module (which is indicated by the namespace PhysicalEntity).

Listing 4.11: Specification of temperature values in bathrooms in the *HAO* module

Class: PhysicalEntity:Bathroom

    SubClassOf:
    rdfs:comment   ' The HAO module re−uses and extends the definition
    of Bathroom that is taken from importing the PEO module.
    If a Bathing action happens in the bathroom, bathroom instances
    are valid (i.e., temperature values can be ignored).
    Also, if a Bathing action happens in a room, this room is
    automatically of type Bathroom. ' @en
        (inv (happensIn) **some** SpatialAction:Bathing)

    rdfs:comment   ' Otherwise, bathroom instances have to satisfy
    the following two requirements:
    (i) Every physical bathroom is related to a structural room in the
    architectural environment with a temperature sensor. Therefore,

> *in the ADO module, an instance has to exist that is related to*
> *the Bathroom in the HAO module by the relation eConnRealizes*
> *(cf. chapter 5 for more details on this relationship).'@en*
>   **or** (EConnEntities : eConnRealizes **only** (PhysicalEntity : Bathroom
>
> *rdfs : comment ' This ADO instance also has to be represented*
> *by a room region in the ACO module, ...'@en*
>   **and** (inv (BuildingStructure : functionalType) **only**
>   (BuildingStructure : Room
> *rdfs : comment ' ...in which the region of a TemperatureSensor*
> *has to be located (i.e., the sensor region is a proper part*
> *of the room region).'@en*
>   **and** (RBO : inverseProperPartOf
>   **some** (ArchitecturalConstruction : TemperatureSensor
>
> *rdfs : comment ' (ii) This temperature sensor of the bathroom*
> *has to measure (normal) temperature values within a certain*
> *range (as no bathing event happens).'@en*
>   **and** (BuildingStructure : sensorValue **some**
>   { '18C−22C ' ^^xsd : string })))))))

Although the *HAO* module provides a home automation specification and it can thus be built on the basis of the modules described above, similar modules may be defined that follow *HAO*'s formalization but provide information for other purposes. For example, a module for *accessibility* could be specified, in which access information for handicapped user groups is provided. In a similar way as the *HAO* module, this accessibility module would be based on physical entities, structural features, and spatial actions.

Table 4.6 gives an overview of the technical details of the *HAO* module. It is applied in the context of the assisted living application discussed and evaluated in section 6.2.2. The performance of the *HAO* module is thus analyzed in terms of its applicability for an assisted living environment.

4.2.4 Ontology Modules for Multimodal and Cognitive Space

Ontology modules that describe spatial information from a multimodal perspective are primarily intended to be used in human-computer interaction applications. Spatial specifications for modalities, such as gesture, tactile sense, or language, can be used for representing spatial information that is relevant for the interaction or understanding

Table 4.6: Details of the Home Automation Ontology Module ADO. Specification overview of the ontology module that defines home-automation-related requirements in assisted living environments.

	Home Automation Ontology Module
Ontology Module	$HAO = \langle C_{HAO}, R_{HAO}, In_{HAO}, A_{HAO}\rangle$
Spatial Perspective	SP_{HAO} : Domain-specific perspective (assisted living environment)
Concepts	C_{HAO} : Object categories in assisted living environments, reusing other modules
Relations	R_{HAO} : Assisted-living-related constraints and new category restrictions
Instances	In_{HAO} : Entities reflecting the objects in the assisted living environment
Axioms	A_{HAO} : Home automation requirements
Purpose and Aim	Description of valid situations in assisted living environments
Requirement	Provide specification in close relation to the application
Example Application	Assisted living environments (see section 6.2)
Development	Ontology categories and relations reused from the modules PEO, SAO, and ADO
Ontology Language	DL (OWL 2), supported by DL reasoners
Expressivity	$\mathcal{SRIQ}(D)$, ~500 categories and ~280 relations
Availability	Available online at `http://www.informatik.uni-bremen.de/~joana/ontology/modSpace/HomeAutomation.owl` (cf. appendix A)

of users. Such specifications naturally depend on the cultural contexts for which they are designed or in which they are applied. In the following, an ontology module that specifies language-specific spatial information is presented.

4.2.4.1 Linguistic Spatial Ontology

The linguistic spatial ontology module specifies the constructions that are used in natural language for describing spatial information. It thus has to formulate semantic spatial categories and roles that can be found in language, and consequently this linguistic perspective is closely related to specific natural languages, e.g., English. For example, the ontology formalizes how paths in motion expressions can be used by defining start, intermediate, and end points of the path. Therefore, the requirements for the linguistic module are to analyze natural language and reflect its use and specification for spatial semantics categories, such as motion paths and their characteristics and constraints. The module's purpose, however, is not only to provide a language-specific categorization of spatial types but also to provide a semantic foundation for spatial interaction systems.

An example for an application that can apply this linguistic module is a spatial assistive system that communicates with its users by using natural language. In the context of navigation-related tasks, for instance, the system has to understand the requested destination linguistically expressed by users in order to provide them with route instructions how they can reach their destinations. For such application tasks, the interaction system can use the linguistic semantic categorizations to understand and generate natural language as well as link the language to the spatial environment [Ross, 2008]. Here, the natural language processing is based on several representational layers, one of which is the linguistic module.

The ontology that is used for the linguistic spatial ontology module introduced in the following is *GUM-Space* [Bateman et al., 2010b], which is the spatial extension of the Generalized Upper Model (GUM) [Bateman et al., 1995]. It has been developed within the projects I1-Ontospace[13] and I5-Diaspace[14], which are part of the Collaborative Research Center on Spatial Cognition (SFB/TR 8), funded by the Deutsche

---

[13]http://www.ontospace.uni-bremen.de (visited on January 12, 2011)

[14]http://www.diaspace.uni-bremen.de (visited on January 12, 2011)

Forschungsgemeinschaft (DFG), located at the universities of Bremen and Freiburg[15]. GUM is an ontological specification for generic semantic principles of natural language, primarily developed to be used in dialogue systems for natural language processing. It is based on findings from linguistic research, empirical linguistic evidence, and lexicogrammatical characteristics. Its structure assigns semantic categories to linguistic units and provides a formalization for their possible appearances and dependencies (i.e., ontological relations and constraints).

The main spatial schemas that can be distinguished in natural language expressions should thus be specified in the ontology module for spatial language. In particular, the following linguistic spatial constructions can be identified: (1) the *trajector*, also called *local/figure object*, *locatum*, *referent*, or *target*, which is the entity whose location or position is described; (2) the *landmark*, also called *reference object*, *ground*, or *relatum*, which is the reference entity in relation to which the location or the motion of the trajector is specified; (3) the *spatial indicator*, which defines a spatial relation or some constraints on spatial properties like the location of the trajector with respect to the landmark; (4) the *motion indicator*, which provides specific spatial motion information; (5) the *path*, which may consist of start, intermediate, and end points and which provides path information for a movement, a re-orientation, or a location of an entity; and (6) the *frame of reference*, which indicates the perspective from which a spatial relation holds and which can be classified as *intrinsic*, *relative*, or *absolute* [Levinson, 2003]. [Talmy, 2006; Zlatev, 2007]

GUM's top-most classification distinguishes three types, namely semantic types of parts in sentences (Element), sentences (Configuration), and sentence combinations (MultiConfiguration). The category Element subsumes single objects or conceptual items, which participate as constituent parts in configurations. Linguistically, instances of Element are expressed by verbal groups, nominal groups, adverbial groups, prepositional phrases, and conjunction groups. The category Configuration subsumes activities or states of affairs, i.e., representations of experience, which are expressed at the level of the clause. The category MultiConfiguration describes a sequence of configurations, as expressed by a clause complex that reflects dependencies between single clauses. Hence, the three subtypes represent three different levels of complexity of entities that

---

[15]http://www.sfbtr8.spatial-cognition.de

are related in various ways to each other, as defined by the relation hierarchy. For example, an Element may act as a participantInConfiguration in a Configuration.

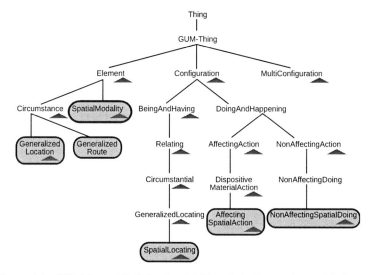

Figure 4.9: *GUM-Space* **Module: Spatial Linguistic Categories and Relations.** Refinement of GUM categories in the *GUM-Space* module are shown in Boxes. Triangles indicate further subcategories.

The linguistic spatial ontology module *GUM-Space* [Bateman et al., 2010b] extends and refines these types by introducing new categories and relations specifically for spatial language, in particular, for German and English. Sentences that provide information about spatial locations (SpatialLocating) or actions (Non-/AffectingSpatialAction) are specified as specific subclasses of Configuration, places (GeneralizedLocation) and locatings (SpatialModality) are specified as subclasses of Element. Their dependencies are constrained by several relations defined in *GUM-Space*. In the following, the main spatial categories, which are shown in figure 4.9, and their relations in the linguistic module are presented. A full report on *GUM-Space* is available in Hois et al. [2009c].

SpatialLocating. The category SpatialLocating is subsumed under GeneralizedLocating, which defines general locations of entities in space, time, or abstract places. As a subconcept of Circumstantial, it defines a circumstantial relationship that relates an

127

entity ('attribute') to its cause ('domain') in space, time, abstract space, or other circumstance-like entities [Bateman et al., 1995]. A SpatialLocating instance specifies static spatial relationships, in which some object is located in space. It is related to an entity that reflects the locatum, an entity that is located with respect to a specific location, the placement. The placement of a SpatialLocating is given by a GeneralizedLocation or GeneralizedRoute (see below). SpatialLocating (just like all subcategories of Configuration) defines a Process that specifies the processInConfiguration. An instantiation of a SpatialLocating is, for instance, the sentence "The book is on the table", in which an entity (the book) is located by means of a static placement" (on the table). Listing 4.12 shows the formalization that specifies the class definition of the SpatialLocating category.

Listing 4.12: Specification of the category SpatialLocating in the *GUM-Space* module

Class: SpatialLocating

  Annotations:
    rdfs:comment ' Any configuration whose function it is
    to locate some physical object in space. Instances must
    be related to one locatum (the located object) and one
    placement (the position of the located object). ' @en

  SubClassOf:
    gum: GeneralizedLocating ,
    placement **min** 1 (GeneralizedLocation **or** GeneralizedRoute) ,
    locatum **exactly** 1 gum: SimpleThing

NonAffectingSpatialDoing and AffectingSpatialAction. Dynamic spatial actions of an entity (actor) that do not affect other entities (actees) are specified by the category NonAffectingSpatialDoing. Spatial actions that do affect other entities are specified by the category AffectingSpatialAction. Both categories reflect spatial information about motion, change in locations, or re-orientations in the same way, except that in the first case moving entities perform the action themselves and in the second case moving entities are forced to move by some other effects. Hence, the primary distinction is that instances of AffectingSpatialAction have to be necessarily related to actees and instances of NonAffectingSpatialDoing have to be necessarily related to actors. Both categories, however, specify further subtypes, which are distinguished according to their type of motion, namely NonAffectingDirectedMotion, AffectingDirectedMotion, NonAffectingOrientationChange, AffectingOrientationChange, NonAffectingSimpleMotion, AffectingSimpleMotion, NonAffectingOrienting, and AffectingOrienting.

DirectedMotion categories specify spatial actions, in which the direction or the path of the motion is given. This information can be expressed in different ways: (i) by a motionDirection depending on the actor, for example, "He walks forward" (non-affecting) and "He rolls the ball downwards" (affecting), (ii) by a direction in case a re-orientation of the motion or of the direction takes place, for example, "They turn left" (non-affecting) and "They carried the boxes uphill" (affecting), or (iii) by a route that may be expressed by the locations source, pathPlacement, pathIndication, or destination (see below), for example, "He went out of the house, through the garden, past the barn, to the gate" (non-affecting) and "He puts the ball in the box" (affecting). Listing 4.13 shows the specification for the AffectingDirectedMotion category that requires either a direction or route to provide the information about places of the motion.

Listing 4.13: Specification of the category AffectingDirectedMotion in the *GUM-Space* module

Class: AffectingDirectedMotion

  Annotations:
    rdfs:comment   'Affecting motion that includes a source, path, destination, and/or pathIndication; or it includes a direction.' @en

  SubClassOf:
    AffectingMotion ,
    (motionDirection **min** 1 GeneralizedLocation)
    or (route **min** 1 GeneralizedRoute)

  DisjointWith:
    AffectingSimpleMotion ,
    AffectingOrientationChange

Instances of OrientationChange specify a change in orientation, which is formalized by an orientationDirection or a route, for example, "He turns away from the window" (non-affecting). The SimpleMotion category specifies action events, in which no clear information about the spatial motion is available, for example, "He swirled the water" (affecting). The Orienting category specifies a spatial configuration, in which an entity is located in space with regard to a certain orientation but without change in orientation, for example, "He pointed the camera at her" (affecting).

GeneralizedLocation and GeneralizedRoute. The categories GeneralizedLocation and GeneralizedRoute are subtypes of GUM's category Circumstance, which provides circumstantial information in configurations that is primarily about time, place, manner, quality, or intensity. Both categories represent the most general constructions to position an entity in space. A GeneralizedLocation specifies a category that represents relative positions of spatially located entities. It is minimally defined by a spatial modality (hasSpatialModality), i.e., the spatial relative relationship between the located and a (potentially undefined) related entity, the relatum. GeneralizedLocation instances are used within spatial configuration as places and parts of routes, for example, "in the warehouse" and "on the right". A GeneralizedRoute represents a route description of static or dynamic spatial configurations. It may define sources, pathIndications, pathPlacements, and destinations that relates the route to its single route elements, that are specified again by GeneralizedLocations. Examples are "from the city to the country" or "across the street past the station".

SpatialModality. Any spatial description contains information concerning the type of relationship being described. This information is typically expressed by a spatial preposition, an adverb, or an adjective. In *GUM-Space*, this crucial spatial expression is represented by the concept SpatialModality. It fills the relation hasSpatialModality of a GeneralizedLocation and represents the type of spatial relationship between a locatum and a relatum in a static configuration, or an actor and a relatum in a dynamic configuration. Its most general distinction is whether a spatial modality reflects distance information between entities (SpatialDistanceModality), functional dependencies between entities (FunctionalSpatialModality), or relative positions between entities (RelativeSpatialModality).

An overview of the *GUM-Space* module is given in table 4.7. Its applicability and use is discussed in section 6.4, in which the module is related to qualitative spatial calculi in order to interpret spatial language. As *GUM-Space* closely reflects spatial language semantics, its evaluation has been carried out consequently by using linguistic techniques for content analyses [Hois, 2010b], described in section 6.4. The module specification is available in OWL and CASL, the first can be used together with DL reasoners in dialogue systems, the second can be used together with more complex representations of spatial information, such as spatial logics, with first-order theorem provers.

4.3 Reasoning with Spatial Ontology Modules

The spatial ontology modules presented above are intended to be used within different spatial assistive systems, such as those presented in chapter 6. The specifications above reflect the representation of certain spatial aspects and reasoning over these aspects and instantiating the representations in an application is technically supported by ontology reasoners. Depending on the specific application which can use one or more of the ontology modules, available reasoners were selected on the basis of their reasoning support. As most of the ontologies are formulated in Description Logics, primarily DL reasoners are used.

The vision application example in section 6.3 applies the DL reasoner Pellet [Sirin et al., 2007], which provides standard ontological reasoning features, such as consistency checking, ontology classification, concept satisfiability, and query answering, which are supported by the OWL API [Horridge and Bechhofer, 2009]. Pellet is based on tableaux algorithms to support reasoning with $\mathcal{SHOIQ}(D_n)$ ontologies. In the vision application, query answering over category specifications in the *PEO* module is particularly relevant, and supported by Pellet.

For the architectural design application (section 6.2), specific region-based reasoning is required. Therefore, the DL reasoner RacerPro [Racer Systems, 2007] is used, which is a tableau-based and algebraic DL reasoner with the query language nRQL, that additionally supports SBox reasoning, a method that implements region-related model checking for RCC-5 and RCC-8. This SBox reasoning can specifically be used and adjusted to the spatial application for analyzing region-based relations and spatial requirements consistency. Furthermore FaCT++[16], a tableaux-based OWL-DL reasoner that is also integrated in Protégé, has been used for reasoning with the home automation scenario presented in section 6.2.2.

In the case of spatial language that is interpreted in the form of spatial calculi (see section 6.4), more complex specifications than DLs are used, and thus a first-order rea-

---

[16]`http://owl.man.ac.uk/factplusplus` (visited on August 05, 2011)

Table 4.7: Details of the Linguistic Spatial Ontology Module *GUM-Space*. Specification overview of the ontology module that defines linguistic constructions in spatial language.

	Linguistic Spatial Ontology Module
Ontology Module	$GUM\text{-}Space$ $=$ $\langle C_{GUM\text{-}Space}, R_{GUM\text{-}Space}, In_{GUM\text{-}Space}, A_{GUM\text{-}Space}\rangle$
Spatial Perspective	$SP_{GUM\text{-}Space}$: Linguistic (modal) perspective
Concepts	$C_{GUM\text{-}Space}$: Semantic categories in natural language expressing spatial constructions
Relations	$R_{GUM\text{-}Space}$: Relations among semantic categories in natural language
Instances	$In_{GUM\text{-}Space}$: Entities reflecting expressions of natural language (e.g., speech or text)
Axioms	$A_{GUM\text{-}Space}$: Dependencies among semantic categories and relations reflecting how natural language is structured and constructed
Purpose and Aim	Language-specific constructions of spatial information
Requirement	Analysis of natural language semantics categories, their characteristics, and constraints
Example Application	Natural language interpretation (see section 6.4)
Development	Reuse of linguistic ontology for natural language (GUM) and reuse of terms from linguistic research for category names [Bateman et al., 2010b]
Ontology Language	DL (OWL 2), supported by DL reasoners; CASL supported, by SPASS reasoner
Expressivity	$\mathcal{SHQ}(D)$, ~280 categories and ~110 relations
Availability	Available online at `http://www.ontospace.uni-bremen.de/ontology/stable/GUM-3-space.owl` (cf. appendix A)

soner is applied. Hence, SPASS [Weidenbach et al., 2009] is used to prove dependencies and requirements of the specification between the spatial language module and the spatial calculus module. SPASS is a first-order theorem prover that is primarily based on resolution extended by ordering, backtracking, rewriting, and further improvements.

As some ontology modules depend on other modules by refining or extending them and as ontologies of the same spatial perspective tend to be closely related with each other, combinations or collections of these modules are likely and they are often relevant for application-specific purposes (see chapter 6). Different technical and semantic methods for combining or relating different modules are thus required and further examined in terms of their applicability and reasoning support in section 5.1.

Moreover, some ontology modules are applied in application contexts that have to deal with either uncertain, imprecise, or unavailable data. In this case, ontology module instantiations may be difficult to apply or they are affected by different types of uncertainties, and the application has to deal with corresponding uncertain reasoning results. Section 5.2 shows the different uncertainty types and ways of dealing with them in spatial applications.

4.4 Chapter Summary

In this chapter, the development of spatial ontology modules was presented and concrete modules were introduced according to the spatial perspectives framework given in the previous chapter. For each ontology module, its purpose and requirements and its implemented specification have been presented. The clear distinction of ontologies according to their spatial perspective not only precisely models conceptual considerations but also supports technical implementation issues.

As the ontology modules reflect spatial information about the same part of the real world, they can be related with each other and combinations of them can be defined. Such combinations are often useful or required for application scenarios, as shown in chapter 6. However, there exist different methods how to combine the spatial modules with each other depending on their ontological specifications. Such methods as well as concrete combinations between spatial ontology modules are presented and discussed in the next chapter together with potential uncertainties in the ontological specifications and instantiations that arise in spatial applications.

5

Combinations and Uncertainties of Ontology Modules

In this chapter, mechanisms for combining different ontologies and integrating different types of uncertainties are discussed and analyzed.

As shown in chapter 3, different ontologies provide diverse types of information about space. However, in application contexts they may need to be integrated or connected with each other. Hence, methods for such integrations and connections are required that address relevant technical and content-related aspects. In the following, techniques for combining multiple ontologies are presented as far as they are used and applicable for the spatial ontologies developed in chapter 4.

In addition to the requirement that multiple ontologies have to be connected with each other, application-specific contexts often have to deal with uncertainty aspects in terms of instantiations or combinations of ontologies. For this purpose, the uncertainty types that can appear in such contexts are investigated, and applicable representation and reasoning strategies for the potential uncertainty types are provided.

5.1 Formalizing Combinations of Modules

As discussed in chapter 4, certain tasks or applications require information from different spatial ontology modules and thus the combination of modules is essential to accomplish these tasks or application aims. In the following, we will analyze technical and content-related aspects of such combinations. In particular, the required formalisms for combining different spatial ontologies are discussed for each combination case. Potential formalisms are introduced and discussed in more detail in order to combine spatial ontology modules. In summary, this section analyzes when, which, and how combination techniques are used for different spatial ontologies and how modularity can support the combination process.

In section 2.4, methods for combining ontologies were introduced and categorized into matching/alignment, merging, extension/refinement, bridging, connection, and blending. Although the methods differ in their applicability and intended purpose, they have in common that they consist of a set of ontology modules and combinations thereof [Serafini and Homola, 2010]. Some of these mechanisms have been applied in the design of modular ontologies for space (section 4.2), namely extension and refinement, matching, and connection. In the following, these techniques, their implementations, and how they are used for combining the spatial ontology modules are introduced.

5.1.1 Extension and Refinement

Extensions and refinements of ontologies use an existing ontology and add new information to it. Either the contents of the ontology can be further *extended* by integrating the categories and relations of the existing ontology into new categories and relations, i.e., the categories and relations of the existing ontology are subsumed under new categories and relations; or the contents of the ontology can be further *refined* by integrating new categories and relations into the categories and relations of the existing ontology, i.e., new categories and relations are subsumed under the categories and relations of the existing ontology. Whereas extensions broaden or add scope and objectives of the *extended ontology*, refinements narrow or define more precisely the scope and objectives of the *refined ontology*.

Figure 5.1 shows an example of the combination of two ontologies by extension and refinement. The category Event subsumes Action and Phenomenon and is part of one ontology module, the category SpatialAction subsumes DoorOpening and RoomEntering and is part of another ontology module. The first categories are more general and broader, the latter are more specific and narrow in their scope and definitions. Integrating the specific with the general categories refines the Action category, and accordingly the more general ontology, by introducing new subcategories; it extends the SpatialAction category, and accordingly the more specific ontology, by subsuming it under Action.

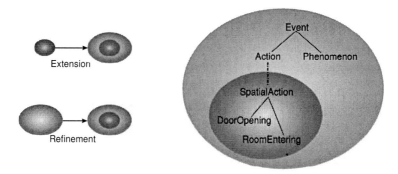

Figure 5.1: Extension and Refinement. Categories particularly for spatial actions defined in the *SAO* module are more specific than actions or events defined in the DnS. Hence, the *SAO* module refines the DnS module and DnS extends the *SAO* module. However, the *SAO* module was initially developed as a refinement of DnS.

On a content-based comparison, an ontology that extends another thus describes more abstract categories and relations whereas an ontology that refines another describes more specific categories. On a technical-based comparison, the refining ontology often has a stronger axiomatization than the extending ontology, which reflects a fine-grained versus a coarse-grained category distinction [Kutz et al., 2010b]. Extensions and refinements may furthermore introduce categories or relations merely for a better integration of one ontology into the other, i.e., an intermediate category or relation may be specified that 'glues' together categories or relations from the extended ontology and refined ontology [McNeill, 2006]. In the example in figure 5.1, for instance, a category subsumed under Action could be defined, which then subsumes SpatialAction.

A specific type of extensions and refinements is *conservativity* [Konev et al., 2009]. If an extension (refinement) is conservative, it does not add new information of what follows from the specification in the extended (refined) ontology. Formally, an ontology O_2 is a conservative extension of O_1 if all assertions made in the language of O_1 that follow from O_2 already follow from O_1. As a consequence, O_1 specifies its vocabulary independent from O_2. With regard to its computational complexity, proving conservativity is decidable for most DLs [Cuenca Grau et al., 2009a], i.e., algorithmic solutions are available, but undecidable for first-order logic and highly expressive DLs [Stuckenschmidt et al., 2009]. As the modular ontologies for spatial information presented in section 4.2 are less expressive, conservativity, if applicable, can be proven, and the complexity of the spatial ontology modules is tractable.

The simplest case of a conservative extension is a *definitorial* extension [Konev et al., 2009]. This type of extension extends only the vocabulary of an ontology by introducing new terms, whose meanings are entirely determined by the axioms of the extending ontology. An example of a definitorial extension was shown in figure 5.1, in which new subcategories are introduced and subsumed by an existing category from the refined ontology module.

In general, extensions and refinements are often-used and well-established techniques for combining ontology modules with each other. They are also easy to apply, e.g., in the case of definitorial extensions. Extensions and refinements have thus been used several times for the specification of the spatial ontology modules: The *SAO* module is a refinement of the DnS ontology (section 4.2.2.1), the *PEO* module is a refinement of the DOLCE-Lite ontology (section 4.2.3.1), the *HAO* module is an extension of the connected *PEO*, *SAO*, and *ADO* modules (section 4.2.3.3), and the *GUM-Space* module is a refinement of the GUM ontology (section 4.2.4.1). In this thesis, both extensions and refinements can be used between any ontological modules, however, it is most likely that modules that comply with an abstract space perspective are used as extensions and modules with qualitative, quantitative, or domain-specific spatial perspectives are used as refinements, which is caused by their detail of specification about the spatial entities they describe (cf. section 3.2).

As most spatial ontology modules are formulated in OWL, extensions and refinements are technically implemented in OWL by using the owl:import statement, which is an inheritance mechanism for OWL ontologies. This allows the integration of an ontology into another ontology, i.e., the importing ontology can reference the specification (e.g., categories and relations) of the imported ontology by its namespace. The importing ontology can then refine or extend the imported ontology and the imported ontology is unaltered. Hence, the importing ontology specifies its own categories, relations, or constraints and it combines them with the imported ontology. An example of this import statement is shown in listing 5.1.

Listing 5.1: Import of DOLCE-Lite in the *PEO* module

```
rdfs:comment   'Namespace declaration in the PEO module:
                PhysicalEntity.owl.' @en
Namespace:  <http://www.informatik.uni−bremen.de/~joana/ontology/
                modSpace/PhysicalEntity.owl#>
Namespace:  dolceLite  <http://www.loa−cnr.it/ontologies/DOLCE−Lite.owl#>
rdfs:comment   '(Omitted further namespace declarations for clarity.)' @en

rdfs:comment   'The PEO module imports the DOLCE−Lite ontology.' @en
Ontology:  <http://www.informatik.uni−bremen.de/~joana/ontology/
                modSpace/PhysicalEntity.owl>
Import:  <http://www.loa−cnr.it/ontologies/DOLCE−Lite.owl>

rdfs:comment   'The relation 'contains' in PEO is subsumed by and refines
                the relation 'generic−location−of' in DOLCE−Lite,
                indicated by the 'dolceLite' namespace.' @en
ObjectProperty:  contains

    Domain:
        PhysicalBuilding  or  PhysicalRoom

    Range:
        PhysicalRoom  or  SpatialEntity

    InverseOf:
        containedIn

    SubPropertyOf:
        dolceLite:generic−location−of
```

Here, the *PEO* module defines its own (empty) namespace and introduces the namespace for the DOLCE-Lite ontology as dolceLite. The DOLCE-Lite is then imported and identified by its URI. Its specification can be referenced and reused by using the namespace dolceLite. For instance, the relation contains in the *PEO* module refines generic-location-of from DOLCE-Lite as it inherits the specification of generic-location-of and adds further constraints about its domain, range, and inverse properties. Thus, it introduces a more fine-grained distinction of generic-location-of relations and narrows the scope of the imported ontology.

A more complex example is shown in listing 5.2 (cf. listing 4.11 for further explanation about the category specification). Here, a refinement of the Bathroom category is specified in the *HAO* module. Most importantly, this category is imported from the *PEO* module and its category definition is refined in the *HAO* module, i.e., a higher axiomatization is provided for the Bathroom category. However, the category definition in the *PEO* module is not affected by this. Moreover, the restrictions on the Bathroom category specify several relations and categories that are imported from other modules or combinations of other modules. The namespaces indicate their imported ontology location accordingly. Also, the *HAO* module imports already combined ontology modules, namely the *EConnEntities* module (cf. section 5.1.3 for more details on this module combination). In the Bathroom example, the *HAO* module makes use of the relation EConnEntities:eConnRealizes from the *EConnEntities* module (defined below in section 5.1.3), which allows a combination of the *PEO*, *ADO*, and *ACO* modules. As the *HAO* module brings together all these different modules, it also widens their scope and definition, i.e., it also extends the other modules.

Listing 5.2: Specification of temperature values in bathrooms in the *HAO* module

```
rdfs:comment   ' Namespace declaration in the HAO module:
               HomeAutomation.owl. ' @en
Namespace:  < http://www.informatik.uni−bremen.de/~joana/ontology/
            modSpace/HomeAutomation.owl#>
Namespace:  SpatialAction  < http://www.informatik.uni−bremen.de/~joana/
            ontology/modSpace/SpatialAction.owl#>
Namespace:  ArchitecturalConstruction  < http://www.informatik.uni−bremen.de/
            ~joana/ontology/modSpace/
            ArchitecturalConstruction.owl#>
Namespace:  RBO  < http://www.informatik.uni−bremen.de/~joana/ontology/
            modSpace/RCC−Ontology.owl#>
Namespace:  PhysicalEntity  < http://www.informatik.uni−bremen.de/~joana/
            ontology/modSpace/PhysicalEntity.owl#>
Namespace:  BuildingStructure  < http://www.informatik.uni−bremen.de/~joana/
            ontology/modSpace/BuildingStructure.owl#>
Namespace:  EConnEntities  < http://www.informatik.uni−bremen.de/~joana/
            ontology/modSpace/EConnEntities.owl#>
rdfs:comment   ' (Further namespace declaration omitted for clarity.) ' @en


rdfs:comment   ' The HAO module imports the SAO module and a
               combination of the PEO, ADO, and ACO modules. ' @en
Ontology:  < http://www.informatik.uni−bremen.de/~joana/ontology/
           modSpace/HomeAutomation.owl >
```

Import : < http ://www. informatik . uni−bremen . de /˜joana / ontology /
modSpace/ SpatialAction . owl >
Import : < http ://www. informatik . uni−bremen . de /˜joana / ontology /
modSpace/ EConnEntities . owl >

rdfs :comment *' The entity Bathroom is further refined in the HAO module.*
In particular , no new Bathroom category is introduced in the
HAO module , but the Bathroom category from the PEO
module (with its namespace PhysicalEntity) is specified
by further contraints in the HAO module. ' @en

Class : PhysicalEntity : Bathroom

rdfs :comment *' Also , the refinement of the Bathroom category in the*
HAO module reuses several relations and categories from
the imported modules , which is indicated by their
different namespaces . ' @en

SubClassOf :
(inv (happensIn) **some** SpatialAction : Bathing)
or (EConnEntities : eConnRealizes **only** (PhysicalEntity : Bathroom
and (inv (BuildingStructure : functionalType) **only**
(BuildingStructure : Room
and (RBO: inverseProperPartOf
some (ArchitecturalConstruction : TemperatureSensor
and (BuildingStructure : sensorValue **some**
{ ' 18C−22C ' ˆˆxsd : string })))))))

Consequently, extensions and refinements are a convenient technique for combining
the spatial ontology modules whenever their specification needs to be specified more
generally with extensions or more precisely with refinements. Extensions primarily
combine ontology modules with different spatial perspectives, as they broaden their
scope and application; refinements particularly support the integration of ontology
modules from the spatial perspective of abstract space. Extensions and refinements
can in principle be used at any level of spatial perspectives as both methods provide
a rather general way of combining ontologies. In fact, they are highly relevant and
often appropriate for the combination of ontology modules exactly because they provide intelligible and easy to apply combination methods. As a result, extensions and
refinements are used in all application examples presented and evaluated in section 6.1.
An overview of extensions and refinements used for the specification of spatial ontology
modules is illustrated in figure 5.2.

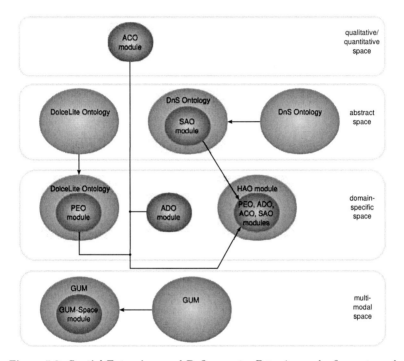

Figure 5.2: Spatial Extensions and Refinements. Extensions and refinements used for the specification of the spatial ontology modules: The *SAO* module refines the DnS ontology, the *PEO* module refines the DOLCE-Lite ontology, the *HAO* module refines and extends the *PEO*, *ADO*, *ACO*, and *SAO* modules, and the *GUM-Space* module refines the GUM ontology.

5.1.2 Matching

Matching provides a method to define so-called *alignments* between categories or relations from two different ontologies. An alignment defines a relation between two categories from different ontologies that are semantically equivalent, i.e., the two categories have the same meaning and share the same set of instances. Matching is the process of identifying alignments between two ontologies. This method does not change any of the aligned ontologies or their specifications. Instead it provides a translation mechanism of parts (categories, relations, or instances) from one ontology into the other. The term ontology matching is equivalent to *ontology mapping*. In a more general sense, matching is also regarded as the identification of alignments between hierarchical or graph-like structures [Shvaiko et al., 2010].

A matching example is shown in figure 5.3. The dashed lines indicate alignments between the categories from two different modules. For instance, Table in one representation is aligned with Table in another ontology, i.e., all instances that are subsumed by Table in the first representation are also subsumed by Table in the second ontology. The example also shows that inheritance relations do not automatically follow from alignments: The sibling categories Table, CoffeeTable, and DiningTable in one module have a different subsumption relation in the other module. In a broader sense, it also shows that alignments do not affect the modules. Although DiningTable is a sibling category of Table in one module, its aligned category is subsumed by the aligned Table in the other module. Aligned categories do not necessarily have the same naming, for instance, IfcFurnitureType is aligned with FurnitureType.[1]

Several approaches can be used to automatically find alignments between ontologies [Euzenat and Shvaiko, 2007]. For instance, correspondences between two ontologies can be identified by syntactic comparisons of category or relation names. For this purpose, the strings of the names are typically normalized to improve comparison results. Also external lexical resources or terminologies can be applied, e.g., to find synonyms or homonyms. Such linguistic methods are, however, limited to lexical comparisons. Other approaches for finding alignments use the structural features of the ontologies, i.e., their axiomatization and category definitions. For instance, mereological characteristics or connectedness of a category can often be compared with categories from

---

[1]In another example, the category Writer might be aligned with Author.

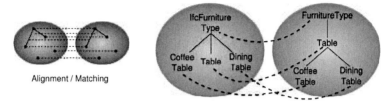

Figure 5.3: Matching. Categories in the *ACO* module are adapted from the IFC specification, i.e., the categories can be aligned with each other. The example also shows that alignments neither have to have the same hierarchical structure nor the same lexical terminology. The aligned definitions match with each other nevertheless.

another ontology. If the category properties are structurally equivalent, alignments may be applied. Alignments can also be identified by corresponding instances by using extensional comparisons between ontologies. Such methods typically use similarity measures to compare two categories from different ontologies that are derived from statistical data or heuristics. Syntactic, structural, and extensional methods can also be used together as a mixed approach for the matching process.[2] Systems are available for automatically finding potential alignments [Euzenat and Shvaiko, 2007; Oltramari et al., 2010], which are based on the different syntactical, structural, and extensional methods, and they are particularly useful for finding correspondences between existing ontologies or for comparing existing sources. Ontology matching (or mapping) can also be used in advance in order to support distributed ontology development and distributed reasoning [Serafini and Tamilin, 2005]. During development, ontology designers are able to develop their own ontology module, and an automatic matching can align these modules with those from other developers. Reasoning is then also distributed over the single modules.

As alignments are used to define equivalence relations between ontology parts (categories or relations), a high number of alignments between ontologies in relation to their overall size (number of categories and relations) indicates a high similarity in terms of their contents. Hence, both ontologies are likely to share the same kind of information of the same domain. In this thesis, alignments are thus mostly applicable between spa-

---

[2]The Ontology Alignment Evaluation Initiative provides a yearly competition to evaluate and compare the different matching algorithms (http://www.ontologymatching.org/evaluation.html, visited on Dec 01, 2010).

tial ontology modules that share the same spatial perspective. As alignments can be defined between general, hierarchical structures, spatial ontology modules can also be aligned with other (non-ontological) representations of space, namely *external sources* for spatial information.

Although automatic matching systems are available, they are less relevant for the spatial ontology modules in section 4.2 as these modules are mostly developed from scratch or on the basis of external sources. Hence, ontology matching is primarily used to align ontology modules with external hierarchical sources [Atencia and Schorlemmer, 2007], and the alignments are manually created during the development phase. Also, due to the development process used for the spatial ontology modules (described in section 4.1.1), each ontology module defines its own scope, purpose, aims, and requirements, and thus similar ontologies (with largely overlapping contents) have not been developed.

The external sources that are used as a basis for specifying certain spatial ontology modules are: the ACO module, which is aligned with the partially hierarchically organized representation of the Industry Foundation Classes (section 4.2.1.2) within the quantitative spatial perspective; the RBO module, which is aligned with the hierarchically structured specification of the RCC-8 relations (section 4.2.1.1) within the qualitative spatial perspective; and the ADO module, which is aligned with conventions of architectural design floor plans (section 4.2.3.2) within the domain-specific perspective. Thus, alignments have not been used between modules from different spatial perspectives but between modules and external sources.

As the external sources are not formally specified, i.e., formulated in any logic or ontology specification language, alignments cannot directly be implemented in OWL or by using other ontological specifications. As an alternative, annotations are used throughout the OWL specifications. In the case of the ACO module, for instance, annotations in OWL are used for referencing the IFC documentation of the aligned (equivalent) categories. An example for this reference is shown in listing 5.3 for the category Building in the ACO module. Thus technically, alignments are manual transformations from another data format into OWL. However, as no new alignments need to be detected between different modules and alignments between modules to external sources have been defined during development, this static transformation is sufficient for the spatial ontology modules.

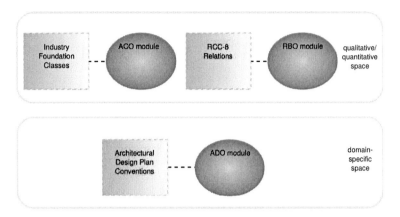

Figure 5.4: Spatial Matchings. Alignments used for the specification of the spatial ontology modules: The *RBO* module corresponds to the RCC-8 relations, the *ACO* module corresponds to the IFC specification, the *ADO* module corresponds to architectural design conventions.

Listing 5.3: Specification of an IFC-aligned category in the *ACO* module

```
Class: Building

rdfs:comment   ' An annotation of the category aligns it with its
                 correspondence (equivalence) in the IFC data model. ' @en
    Annotations:
        rdfs:seeAlso   ' http://www.iai-tech.org/ifc/IFC2x4/alpha/html/
                         ifcproductextension/lexical/ifcbuilding.htm ' @en

    SubClassOf:
        IFCArchitecturalEntity ,
        contains some DistributionElement ,
        contains some StructuralBuildingElement
```

In the spatial ontology framework, alignments are particularly relevant for combining spatial ontology modules with external sources. As the alignments are interpreted solely as equivalence relations, their use is comprehensible and easy to apply. However, the transformed ontological representation of the external sources contains further information, e.g., the example shown in figure 5.3 adds new hierarchical information and axiomatization to the specification and it simplifies category names and relations.

146

Matching with external sources is also important for combining spatial ontology modules with other applications or systems that use these external sources. This also provides a method to integrate the ontological specification into non-ontological specifications or data formats of other systems, e.g., a translation to different formats within a specific system (e.g., IFC and CAAD). An application example is presented in section 6.2. In the broader sense, this technique also supports the integration of different systems or the alignment of different external sources with the ontological module as a reference. An overview of the alignments used for the spatial ontology modules is shown in figure 5.4.

5.1.3 Connection

A *connection* between two ontologies provides a method to combine ontology modules that are rather different and heterogeneous in their specification and contents. The connections define link relations between parts from the different module specifications. These links reflect associations or similarities between categories and relations from the different ontologies with regard to their overall meaning within the specification. The link relations together build an *interface ontology* that reconciles one ontology module with the other. Both modules, however, are kept entirely disjoint from each other and their definitions remain unaltered. The interface ontology can particularly be used to generate an overall ontology module entailing the two input ontologies and the link relations. In contrast to alignments, link relations do not define equivalence relations. Instead, link relations are counterpart relations between two ontologies that relate two functionally related parts. For instance, two categories from different modules may not share the same instances or the same structural definition but may be related by a link relation.

Figure 5.5 illustrates an example of the connection of two ontology modules. Both modules are connected by a set of link relations from one module to the other and vice versa. The example shows the different interpretations of the category Door in different ontology modules. One module specifies a Door by means of its metrical properties, such as length and height, and of specific subtypes of door, such as RevolvingDoor, SwingDoor, and SlidingDoor. The other module specifies a Door by means of its functional characteristics, such as its entrance function to buildings or building parts. However, a door instance can be instantiated in both modules by their respective category definitions for

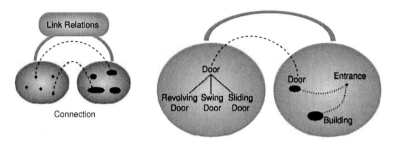

Figure 5.5: Connection. Categories in the *ACO* module are connected with categories in the *ADO* module. The example shows the distinction between the Door category in both modules. The *ACO* module defines a Door to have certain metric properties and to have specific-door subcategories, whereas the *ADO* module defines a Door to provide the function of an entrance to some other category.

Door, i.e., both modules can contain an instance that reflects the same entity in reality, and a link relation between these instances based on the connection of the modules can reflect this information.

Hence, connections can identify link relations particularly between thematically different modules that describe a domain from different perspectives but partially specify the same entities. The contents of the connected modules can highly differ, neither their ontological axiomatizations nor scope and aims nor the granularity of their definitions have to be related with each other. For using module connections not only the contents can be rather disjoint, but also the ontology languages in which the modules are defined can be different. Connections can thus deal with heterogeneous specifications [Kutz et al., 2010b].

A specific form of module connections is provided by $\mathcal{E}$-connections. This formalism provides the link relations as loose couplings that can relate different logics and ontologies [Kutz et al., 2004]. Such connections are intended to be defined manually, e.g., by ontology designers or human experts, potentially during design time. However, automatic extractions of submodules from a single ontological specification are available [Cuenca Grau et al., 2005]. $\mathcal{E}$-connections preserve decidability if the connected modules are formulated in a decidable ontology language [Kutz, 2004], and they are thus particularly interesting for combining logically heterogeneous specifications. $\mathcal{E}$-connections can accordingly be applied on ontologies formulated in OWL and its fragments [Cuenca Grau et al., 2009b], i.e., for every ontology with the expressiveness of $\mathcal{SROIQ}$ or less the connection consequently stays decidable.

The link relations of $\mathcal{E}$-connections are interpreted as a binary directed relation between two categories of different ontology modules. They can thus be seen as ObjectProperties across modules with domain and range. Specific $\mathcal{E}$-connections are then axiomatized as a directed relation from one module into the other (and potentially their inverse relations). As dedicated OWL reasoners for $\mathcal{E}$-connections are not available at the moment,[3] we realize them by using ObjectProperties in a way that allows a complete encoding of the semantics of $\mathcal{E}$-connections into OWL DL (cf. also Cuenca Grau et al. [2006]): first, the separation of the different ontology modules is enforced by defining their top nodes to be disjoint; second, the link relations are restricted in their domain and range definitions so that they relate parts from one module with the other; third, the link relations are used for connecting categories from both modules by using existential and universal operators.

In this thesis, $\mathcal{E}$-connections are applied primarily for connecting thematically disjoint spatial modules, i.e., modules with different spatial perspectives or different aims and scope. For the specification of the spatial ontology modules (section 4.2), $\mathcal{E}$-connections are used to combine the *ACO* module with the *ADO* module resulting in the *EConnArchitecture* interface ontology module, the *EConnArchitecture* module with the *PEO* module resulting in the *EConnEntities* interface ontology module, and the *EConnEntities* module with the *SAO* module resulting in their use in the *HAO* module. Also, $\mathcal{E}$-connections are used for combining *GUM-Space* with spatial calculi, presented in more detail in section 6.4.

Listing 5.4 shows an OWL example of the interface ontology module *EConnArchitecture* that results from connecting the *ACO* and the *ADO* modules. The two modules are enforced to be separate by defining their top nodes to be disjoint. A link relation eConnComposes is specified as an ObjectProperty that connects categories from the *ACO* module (domain: IFCArchitecturalEntity) with categories from the *ADO* module (range: BuildingStructure). By constraining this link relation over specific subcategories from the two modules, the link relationship between the two modules is further defined. For instance, the links between instances of the categories Building from both modules are limited.

---

[3]The Swoop implementation of $\mathcal{E}$-connections in Cuenca Grau et al. [2009b] is still in development.

Listing 5.4: Specification excerpt of the *EConnArchitecture* module

```
rdfs:comment   ' E-Connection interface ontology module that connects
               the ACO (ArchitecturalConstruction.owl) and
               the ADO (BuildingStructure.owl) modules. ' @en

Namespace:  < http://www.informatik.uni-bremen.de/~joana/ontology/
            modSpace/EConnArchStructure.owl#>
Namespace:  BuildingStructure  < http://www.informatik.uni-bremen.de/~joana/
                               ontology/modSpace/BuildingStructure.owl#>
Namespace:  ArchitecturalConstruction  < http://www.informatik.uni-bremen.de/
                               ~joana/ontology/modSpace
                               /ArchitecturalConstruction.owl#>

Ontology:  < http://www.informatik.uni-bremen.de/~joana/ontology/
           modSpace/EConnArchStructure.owl >
Import:  < http://www.informatik.uni-bremen.de/~joana/ontology/
         modSpace/ArchitecturalConstruction.owl >
Import:  < http://www.informatik.uni-bremen.de/~joana/ontology/
         modSpace/BuildingStructure.owl >


rdfs:comment   ' Top nodes from ACO (IFCArchitecturalEntity) and
               ADO (ArchitecturalStructure) are made disjoint. ' @en

Class:  ArchitecturalConstruction:IFCArchitecturalEntity

   DisjointWith:
      BuildingStructure:ArchitecturalStructure


rdfs:comment   ' A link relation (eConnComposes) is defined between
               a category from the ACO module (IFCArchitecturalEntity)
               and a category from the ADO module (BuildingStructure). ' @en

ObjectProperty:  eConnComposes

   Characteristics:
      Irreflexive

   Domain:
      ArchitecturalConstruction:IFCArchitecturalEntity

   Range:
```

BuildingStructure : BuildingStructure

InverseOf :
eConnComposedOf

rdfs : comment   *' A metrically defined category Building in the ACO modules*
is connected with the functionally defined category Building
in the ADO module. Every instance of an ACO building can
only be linked to one instance of an ADO building . ' @en

Class : ArchitecturalConstruction : Building

SubClassOf :
eConnComposes **exactly** 1 BuildingStructure : Building

In the case of spatial modules, link relations are relevant to connect, for instance, quantitative and domain-specific types of spatial information that reflect the same reality but from entirely different perspectives. The link relations are then specified by a human expert (the architect) in order to identify the relationships between thematically different modules. An overall connection of these thematic modules is achieved by $\mathcal{E}$-connecting counterpart categories with appropriate linking axioms (universal or existential restrictions). Section 6.2 discusses and evaluates the application of the connected ontology modules. The more specific case of connecting *GUM-Space* with other modules is introduced and discussed in section 6.4. An overview of the $\mathcal{E}$-connections used in the spatial ontology module specifications is illustrated in figure 5.6.

5.1.4   Summary of Ontology Combinations

The different techniques for combining ontology modules not only provide different theoretical methods but also thematically different inter-module relations, i.e., their (syntactical) specifications as well as their (semantical) interpretations for module combinations vary. With regard to the spatial ontology framework, the different spatial perspectives and their ontology modules (presented in chapter 4.2) are combined in such different ways, in particular, by applying refinements and extensions, mappings and alignments, and connections and link relations. The selection of these methods is based on the thematically different perspectives by which each ontology module defines aspects of spatial informations.

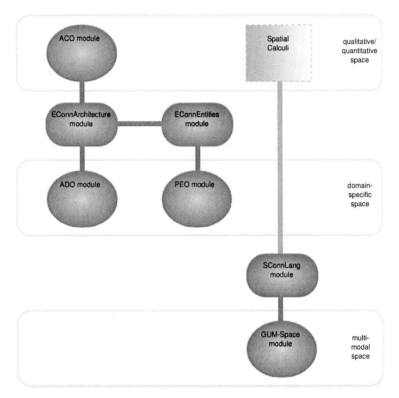

Figure 5.6: Spatial Connections. $\mathcal{E}$-connections used for the specification of the spatial ontology modules. The *EConnArchitecture* module connects the *ACO* and *ADO* modules, the *EConnEntities* module connects the *EConnArchitecture* and *PEO* modules, and the *SConnLang* connects the *GUM-Space* module and spatial calculi.

Using a specification of combined spatial ontology modules provides the following benefits for the overall spatial representation: it reflects interconnected modules that are distinct from each other in their specific spatial description that nevertheless reflect the same 'space'; it thus keeps the different perspectives of the modules distinct and it allows their combinations, exchange of knowledge, and comparisons; it also allows applications to select only relevant modules with combinations between them instead of using one monolithic specification; thus it supports the reuse of smaller modules in different applications; finally, 'contradicting' definitions have no effect in combined ontology module specifications (cf. section 6.2 for an example of disjoint category definitions that co-exist in different modules).

Although the combination methods can generally be used for more than two ontology modules, i.e., multiple mappings or connections are available [Kutz et al., 2010b], only combinations of two spatial ontology modules at a time are specified primarily because of clarity reasons. The spatial framework also does not require the combination of multiple modules, instead mixtures or embedded combinations have been used. For instance, the *HAO* module is a refinement of the *SAO* module and it extends the *EConnEntities* module, which consists of connected other spatial modules.

More combinations techniques are available (discussed in section 2.4), which have not been used in formalizing the spatial ontology modules. In particular, merging, bridging, and blending have not been discussed in this section. Merging was not applicable for the spatial ontology modules, as this technique is commonly more useful when pre-existing modules need to be integrated into each other. As the spatial modules have been designed in a way that clearly separated their purpose, aims, scope, and requirements, no spatial modules with overlapping contents needed to be merged. Bridging was also not relevant for the spatial ontology modules, as there are no scenarios in which source and target ontology modules have to be distinguished. Although connections, for instance, provide a directionality in their link relations, the direction of the information flow is not essential to the combination of spatial ontology modules. Preliminary work has been presented in Hois et al. [2010], in which ontological blending is used for metaphorical definitions and assimilations of spatial categories. Further research, however, is necessary for a more elaborate specification and interpretation for creating blend ontologies.

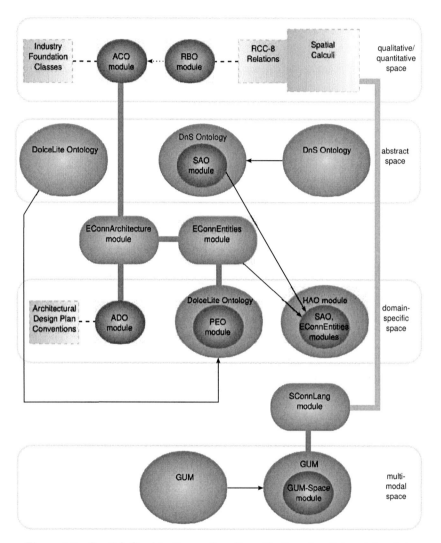

Figure 5.7: Spatial Combinations. Overall specification of spatial modules, their perspectives, and combinations. Arrows indicate refinements, dotted arrows indicate reuse, dashed lines indicate alignments, wide lines indicate link relations.

In summary, refinements and extensions are used for granularity and perspective distinctions in describing spatial information as well as for reusing foundational and abstract spatial ontological specifications. Matching and alignments are applied when external sources are used for spatial module specifications. Connections support the combinations of those modules that have highly distinct perspectives, aims, and scope on the spatial domain. In terms of their technical use, all of these combination methods are applicable and specifiable for OWL ontology modules and all of them are supported by OWL reasoners. In chapter 6, the different formalisms for combining ontology modules and their applicability in different use cases is presented and evaluated. Figure 5.7 illustrates the different spatial modules, their spatial perspectives, and their combinations; it also shows where modules have been simply reused and imported by other modules.

5.2 Specifying Uncertainties for Ontology Modules

As discussed in section 2.5, lack of knowledge, conflicting knowledge, ambiguity, belief, and other unknown facts can cause uncertainties in applications, domains, and knowledge bases, which are modeled by ontological representations in this thesis. Hence, the spatial ontology modules developed in this thesis and intended to be used and applied in spatial assistive systems can be affected by uncertainties. However, not every single application should get its individually developed uncertainty solution for reasons of generality, thus a general foundation for integrating uncertainties in ontological representations and their consequences are analyzed in this section.

We analyze in more detail how uncertainties can affect an ontological representation. First, the following section discusses general evidence of uncertainties and their impact on ontological representations. Briefly, related approaches are summarized and uncertainties in well-established ontologies are discussed, including 'uncertainty' as an ontological category. As a consequence, a semantic analysis for uncertainties in ontologies is investigated in this thesis. The resulting uncertainty formalizations for ontology modules are based on the available ontology constructors and on their chance of being influenced by uncertainty types.

5.2.1 Relevance of Uncertainties in and across Modules

Zimmermann [2000] argues that using only one uncertainty method does not support all
the different kinds of uncertainty information that occur in applications and domains:

> Research is generally done in the frameworks of these axioms and it is
> very seldom investigated which of the theories is adequate to model a spe-
> cific instant or context. It is argued in this contribution that the modeling
> of uncertainty should not be done context free, but that [...] the entire
> information flow from the phenomenon via the uncertainty theory to the
> observer has to be consistent with respect to quality and quantity of infor-
> mation. [Zimmermann, 2000]

Hence, the uncertainty method used in a specific situation should be adequate
and appropriate for the requirements of the domain or application. The uncertainty
method that is used for a specific task, domain modeling, or decision problem has to
take into account the cause as well as the interpretation of the uncertainty, available
information, the information type that is required by the method, and the format how
the uncertainty is to be presented to a final observer (a human decision maker or an
agent). In the same way, uncertainties in ontological specifications have to take into
account the suitability of different uncertainty methods and their applicability.

However, ontologies are not supposed to directly represent uncertainties, as their
specifications of a domain are typically strict and well-defined (cf. section 2.2). As soon
as an ontology is used as an instantiation in a system, however, different types of uncertainties arise. In the case of spatial systems, these uncertainties can be caused by lack
of knowledge about the environment, unreliable sources of information, shortcomings
in sensorimotor data, unknown or unexpected results of actions, or unknown intentions
of other communication partners. Even though the ontology itself may not be affected
by uncertainties, a system's instantiation of it is. Facts of a domain as specified by
an ontology can either be true or false, and a system may be uncertain as to which
is the case. These uncertainties can be modelled by specific syntax and semantics in
ontologies [Hois, 2009].

Foundational ontologies, such as DOLCE [Masolo et al., 2003], do often not have
any categories or relations that reflect or take into account uncertain information nor do

they specify uncertainty as a category type. SUMO [Niles and Pease, 2001], for instance, only provides a relation ProbabilityRelation that assigns a percentage to the probability of an event. Hence, this category does not fully represent potential uncertainties in a domain and it is only available for one category, namely the category Event. Also, reasoning with these probabilities is not available and their interpretations are not further examined. The Cyc[4] ontology, for instance, specifies types of uncertainty only as a feeling of being unsure about something and probability in order to assign values of likelihood to other categories. A clear specification or interpretation of these two uncertainty categories or a systematic integration of uncertainties are not available.

In contrast to an ontological representation of uncertainties, i.e., an ontological analysis if, where, and which uncertainties have to be modeled by an ontological specification, technical aspects for reflecting uncertain information of different types in ontologies have been investigated in more detail (cf. also table 5.1). Representations for uncertainties in ontologies have recently been developed by enhancing the web ontology language OWL. The language has been extended with Bayesian networks [Ding et al., 2006], fuzzy logic [Stoilos et al., 2005], and probabilities [Costa and Laskey, 2006]. Such approaches allow one specific type of uncertainty to be represented in any ontological construction.

Ontological combinations that relate categories and relations of different ontology modules have also been technically extended with uncertainties. Several solutions for this problem were proposed, including Bayesian networks [Mitra et al., 2005] and probabilities [Calí et al., 2007]. In general, such approaches determine uncertain mappings either based on categories or instances. Although these approaches provide uncertain definitions in ontologies, they focus mainly on questions of complexity and expressivity of certain logics. Such approaches define either uncertainties within an ontology or across different ontologies.

Ontological representations have also been modified by integrating vague categorizations. This vagueness is, however, attributed to linguistic underspecifications and vague expressions in natural language [Bennett, 2005], which is different from defining uncertainties in a domain. Such methods analyze specific linguistic aspects of vagueness, not related to a complete analysis of possible uncertainties in domain ontologies

---

[4]http://www.cyc.com (visited on July 05, 2010)

in general. Although vagueness is thus outside the scope of uncertainties in ontologies, it plays a role for interpreting linguistic categories in context.

Given the formal definition of an ontology by Guarino [1998] that an ontological representation should resemble a certain reality as accurately as possible, an integration of uncertainties has to follow an analysis of representing and interpreting adequate uncertainty types. So far it has not been studied in detail which kinds of uncertainties are appropriate and reasonable in an ontological specification. From a semantic point of view, it is thus necessary to analyze which ontological domain definitions may be affected by uncertainties and what their interpretations, i.e., their uncertainty types, are. From a technical point of view, uncertainties can occur only in correspondence with the formulated ontological structures, i.e., the taxonomy of categories, the relations between these categories, their axiomatizations, or their instantiations. As a consequence, the following analysis takes into account the potential occurrences of uncertainties in ontology constructions and their semantic interpretation.

For this purpose, potential occurrences of uncertainties in ontologies are studied with regard to their meaning and interpretation. As uncertainties are mostly caused by lack of knowledge and unknown facts in agent-based uses (instantiations) of an ontology, we thus analyze when and how uncertainties may affect the ontological specifications and which different types of uncertainties are appropriate for reflecting them. The resulting structuring of uncertainties in ontological representations differentiates meanings of uncertainties according to their appearances in the ontological specifications. Thus, it not only reduces and simplifies complexity, as it narrows down the use of uncertainties to appropriate cases, it can also guide ontology engineers in their development of ontologies, as it indicates potential occurrences and uncertainty types in domain categorizations. We thus characterize possible uncertainties in domains and their interpretations, distinguish the different kinds of uncertainties arising, and discuss possible formalizations.

Several methods for representing and reasoning with uncertainties have been studied (cf. section 2.5). These methods differ in their semantics and applicability as well as their reasoning techniques. They can mainly be classified by their uncertainty representation and cause of the uncertainty. Hence, we analyze the possibility of uncertainty types occurring in ontological constructions with respect to the different uncertainty theories and their respective semantics, i.e., we analyze which ontological constructions are affected by which types of uncertainty. If the constructions are affected by uncertainties, a representation is provided that accordingly reflects these types of uncertainty in an adequate way.

As in this thesis the ontologies are developed primarily in OWL and accordingly by using the DL constructors, the representation of uncertainties in ontologies is thus demonstrated by refining syntax and semantics for DL ontologies. Each constructor used for formulating DL ontologies (in $\mathcal{SROIQ}$) is examined according to the different types, causes, and occurrences of uncertainties. This is done closely in relation to the interpretation of the constructors, i.e., which semantic information they reflect in an ontological specification. The syntax available for the individual constructors was introduced in table 2.1 and 2.2. In the following, these constructors are analyzed with respect to their potential uncertainty within and across ontology modules.

An overview of possible uncertainty types and related work is shown in table 5.1. Several approaches have been proposed that integrate probabilistic uncertainty, belief, possibilistic uncertainty, vagueness, or similarity in DLs and OWL. They demonstrate recent responses to the need for extending ontologies with uncertainties, in particular, for combining uncertainties with ontology languages for the Semantic Web [Lukasiewicz and Straccia, 2008]. The methods summarized in table 5.1 provide ways for technically using specific types of uncertainties as part of the ontology language OWL. However, they do not investigate whether and which types of uncertainties should be specifiable as part of an ontology under the conditions of formal ontology, i.e., the adequate modeling of a certain reality as closely to this reality as possible.

5.2.2 Uncertainties within Ontology Modules

An ontology module provides information about categories, relations, and axioms, i.e., a description of entities and their characteristics and dependencies of a specific domain. These representations provide the structure for a knowledge base of a system, and its instantiations reflect application-specific situations. As discussed above, due to lack of knowledge, conflicting knowledge, ambiguity, belief, and unknown facts, these representations can be affected by uncertainties in different ways, either with regard to category, relation, or axiom constructors.

	Short Description	Existing Approaches for OWL
Probability Theory	likelihood degree for statements	BayesOWL [Ding et al., 2006]; probabilistic ontology mapping [Mitra et al., 2004; Pan et al., 2005; Tang et al., 2006]; PR-OWL [Costa et al., 2008]; OntoBayes [Yang and Calmet, 2005]; Pronto [Klinov, 2008]
Evidence Theory	belief about statements	BeliefOWL [Essaid and Yaghlane, 2009]; data fusion [Bellenger and Gatepaille, 2010]; ontology mapping [Nagy et al., 2007; Laamari and Ben Yaghlane, 2007]
Possibility Theory	possibility and necessity of statements	PossDL [Qi et al., 2010]
Fuzzy Set Theory	vagueness degree of statements	fuzzy OWL ontology specification [Bobillo and Straccia, 2009]; fuzzy OWL [Calegari and Ciucci, 2007; Stoilos et al., 2010]; fuzzy mapping [Buche et al., 2008]
Similarity Theory	resemblance between statements	similarity ontology mapping [Maedche and Staab, 2002; Araújo and Pinto, 2008]; similarity in and across ontologies [Ehrig et al., 2005]; Sim-DL [Janowicz, 2006]; similarity ontology alignments [Euzenat and Valtchev, 2004]

Table 5.1: Uncertainties in Ontologies. Uncertainty methods and related approaches for integrating the methods in OWL.

5.2.2.1 Category Constructors

As introduced in section 2.2, categories in an ontology define the entities with regard to their unique characteristics and with regard to their distinction from other categories by grouping them in a hierarchical way. Categories specify properties and constraints that have to be satisfied by the instances of the respective categories. For instance, categories in the *ADO* module are Wall, Window, or Building. Categories thus classify the domain into distinct groups that share the same semantics, and hence they structure the domain by means of the instances occurring in the domain. In a classical ontology, all categories are meant to be well-defined, i.e., their descriptions should be as exhaustively as possible to describe the entities. In particular, it should be feasible to distinguish whether an instance of the domain is a member of a specific category.

Accordingly, there is no indication to use uncertainties with respect to category definitions, and accordingly with the DL atomic concept constructor. The constructor C for atomic concepts, and similarly the construction for general concept inclusions $C_1 \sqsubseteq C_2$, should thus be strict and unaffected by uncertainties. For instance, it should be clear whether the category TemperatureSensor is a subcategory of Sensor given the ontological domain specification.

Nevertheless, if required by a domain specification, C_1 may only be partially entailed in C_2, i.e., not all of C_1's instances should also be instances of C_2. In this case, C_1 can be defined by a union of two categories, C_{11} and C_{12}, of which only one is a subcategory of C_2. For example, the category TemperatureMeasureEntity can be defined as the union of categories Thermometer and TemperatureSensor, in which case only TemperatureSensor is specified as a subcategory of Sensor. This representation reflects the hierarchical and structural dependencies across categories, and it should not be confused with the degree to which an instance might by classified as a specific category. For instance, a concrete instance *room*1 classified as Room may also be classified as an instance of Office but only with a certain degree. However, this does not reflect uncertainties of category definitions but uncertainties that arise from the instantiation of categories, as described below in section 5.2.3.

Another example that argues for uncertainties in category constructions is discussed in Holi and Hyvönen [2005], where geographical areas are used as a motivating scenario. The example illustrates that Lapland together with Finland, Sweden, and Norway cannot

be represented by using only crisp ontology definitions, because the geographical area
Lapland overlaps with the countries Finland, Sweden, and Norway, i.e., the individual
instances cannot clearly be instantiated as the type geographical area without allowing
uncertainties. Also, the geographical area Russia being part of both Asia and Europe indicates that geographical areas cannot clearly be distinguished. However, the
examples rather imply the way an ontological specification should model information
about geographical areas and countries and how to distinguish these types. For instance, if Asia and Europe are defined by the category GeographicalArea and Russia is
defined by the category Country, a relation between countries and geographical areas
can easily solve the location of Russia. Such ontological modeling decisions can best be
resolved by using upper-level classifications of general ontological types, e.g., reusing
DOLCE's classification, and do not imply that category constructions need to integrate
uncertainty types.

Note that one may also argue for the definition of fuzzy categories in an ontology.
This fuzziness is typically caused by underspecified linguistic terminology [Lukasiewicz
and Straccia, 2008]. A library, for instance, is not defined by an exact number of books,
i.e., by a crisp definition. As such vague expressions describe linguistic concepts, they
have to be determined by contextual or real world aspects individually in a specific
situation (e.g., the possibility to read and borrow books) instead of specifying uncertain
category definitions in ontologies. The underspecified linguistic assignments can thus
be interpreted with separate ontological layers, as discussed in [Bateman et al., 2007b].
The connection between these layers may, however, be affected by uncertainties, i.e.,
uncertainties across ontologies (see section 5.2.4).

Unions and intersections of categories can be used for constraints or relations among
several categories. The domain or range of a property, for instance, may be assigned to a
union of categories. For example, the union Wall ⊔ Window ⊔ Door fills the range of the
relation border of Room in an existential restriction $\exists R.C$ (someValuesFrom construction)
in the ADO module. Union and intersection constructions allow flexible relationships
in category restrictions and are thus not affected by types of uncertainties. They work
merely as operators for other constructions. Similarly, the negation constructor $\neg C$
of categories is strict. They negate all category definitions and they are also used
particularly for defining restrictions of category or relation definitions. A category
Table and its negation $\neg$Table have their crisp semantics. Specific attributes determine

the meaning and behavior of Table and its negation accordingly. This differs, however, from an instantiation, i.e., an actual entity in the world. This entity may or may not be of the type Table, specified by the instance constructor (cf. section 5.2.3).

Equivalent categories are typically used to relate different ontologies. We will discuss this construction in more detail in the next section, as it aims at defining relations across different ontologies. Other use of equivalent categories as a logical operator is clearly not affected by uncertainties. Conjunction and disjunction, which combine category definitions, are likewise not influenced by types of uncertainties.

5.2.2.2   Relation Constructors

Relation constructors define relationships between categories and characteristics (properties) of categories. Modeled as ObjectProperties in OWL, they define the domain and range of categories that are related with each other by the relation. Examples of relation definitions in the ADO module are containedIn, adjacentTo, or sensorRange. Such relations can also be classified according to their types, e.g., whether they define a quality or dependency among categories [Masolo et al., 2003]. Thus, they specify constraints necessary for specifying the characteristics of categories. Although it may be unknown whether a specific relation holds for a certain category or instance, the definition of the relation is crisp and not further affected by uncertainty types, because the property definition itself provides a potential relationship with its respective domain and range. The relation (property) hierarchy of an ontology is thus well-defined and also not influenced by any uncertainty. For example, the property containedIn defines the containment relationship between Room and Building. A specific Room office1 can then be containedIn a Building officeBuilding1. Whether this particular instantiation of the property containedIn, namely containedIn(office1,officeBuilding1), is actually true, is irrelevant for the definition of the relation itself, namely that Room and Building are related by this relation per definition.

The transitive role $^{(+)}R$ (transitiveProperty) and inverse role R^- (inverseProperty) constructions define a property to be transitive, e.g., containedIn, or inverse, e.g., neighbor. These logical aspects may either hold for a specific relation or not and are thus not affected by uncertainties. For example, a relation nearTo may define a close distance between two categories and within certain limits may even be considered as transitive to some extent. But given a sequence of instances which are pairwise nearTo each other, it is difficult to say whether the relation nearTo between the first and the last instance still holds. This, however, does not indicate that transitiveProperty has to provide a certain type of uncertainty. Rather, it indicates that the relation nearTo cannot be specified as being transitive in terms of this interpretation.

Similarly, the constructions that provide primarily logical aspects for relation specifications do not integrate types of uncertainty. In particular, the constructors for defining reflexivity, irreflexivity, symmetry, antisymmetry, negation, disjointness, and complex role inclusion may hold between ontological relations (properties) or not [Lukasiewicz and Straccia, 2008]. For instance, the RBO module specifies the relation equalTo as reflexive in alignment with the RCC-8 formalization. Hence, the reflexive characteristic of the relation is crisp, and generally, a reflexive relation does not imply any type of uncertainty.

5.2.2.3 Category Restriction Constructors

Within an ontology, not only category definitions and relation definitions can be specified, but also constraints between them. These are given by the value restriction constructor $\forall R.C$ and the existential restriction constructor $\exists R.C$, i.e., the allValuesFrom and someValuesFrom constructors in OWL, as well as by qualified number restrictions $\geq nR.C$, $\leq nR.C$, and $= nR.C$, i.e., the cardinality constraints. allValuesFrom and someValuesFrom constrain the range of a relation to be of a specific category type. The choice between these two constraints allows a flexible definition within an ontology. Uncertainties in terms of likelihood, for instance, can be covered by someValuesFrom, and additional uncertainties are thus not required with regard to these constraints. For instance, the category Forest can be specified as consisting of at least some (someValuesFrom) different plants or animals, regardless of their actual types or number.

Cardinality restrictions specify the number of relations of a category. In OWL, positive integer values can be assigned to the constraints maxCardinality, minCardinality, and cardinality in order to define the necessary number of relations. This number can thus define how many relations are required at least, at most, in an interval from minimum to maximum, or with a specific number, which allows a flexible mechanism for constraining relations. Just as for value-related constraints, there are no indications for uncertainties for cardinality restrictions. The category Building in the ADO module, for instance, may have a certain number of contains relations to Rooms, ranging from 0 to n. In fact, this interval can be used to model variations in domains and incomplete information. Together with conjunctions and disjunctions, category restrictions are already powerful enough to model domains that cannot be clearly defined in terms of real-world situations. For example, the *PEO* module that is used together with uncertainties in active vision, makes use of these constructions to model categories with less clear boundaries. Here, the category Kitchen can contain a Refrigerator or an Oven, but even without them an instance can still be a Kitchen (see section 6.3).

5.2.3 Uncertainties in Instance Constructors

Instantiations of categories and relations represent a concrete situation structured by the ontological specification. Instances of categories are defined by type, instances of relations are defined by property in OWL. Instantiations reflect the entities that occur in an actual situation or application and that are known to an agent or system. For example, the entities that are part of a building in the *ADO* module can be instantiated based on a given architectural floor plan. But also the entities (e.g., specified by categories in the *HAO* module) that are present in an indoor assisted living system can be instantiated based on monitoring devices of the system. The instantiated information in this system typically relies on sensor data or pre-processing applications that cause potential uncertainties due to errors during sensing or processing.

Table 5.2: Uncertainties affecting Ontology Instance Constructors within Modules. DL instance constructors that are affected by uncertainties within an ontological specification.

Syntax	Semantics	Uncertainty Type	Example
$a : A$	$a \in A^{\Im}$	probability	x is of type DetachableObject with a probability of .8
		belief	The belief that x is of type CopyRoom is .9
$\langle a,b \rangle : R$	$a,b \in A^{\Im} \wedge \langle a,b \rangle \in R^{\Im}$	probability	The probability of a being of sensorType b is .2
		belief	The belief of a being locatedIn b is .8

Both instantiation definitions for categories and relations are thus potentially affected by uncertainties. In particular, a system that instantiates an ontology module in order to represent its environment or situation, can be influenced by various kinds of uncertainties: Input data of a system is vulnerable to inaccuracy, incompleteness, ambiguity, and incorrectness, because of noise, unreliable sources, or limitations of a system's sensorimotor capabilities. Assumptions or conclusions that are drawn may thus turn out to be wrong and lead to additional errors. For example, a spatially-aware system has to classify perceived entities on the basis of its sensory input, which is not only affected by noise but also relies on results from a visual recognition system, as described in section 6.3, and consequently, the classification process is affected by uncertainties. In general, complete knowledge of a domain is often not fully available in natural environments (open world assumption) and an object's type can often not be determined without difficulties. Thus uncertainties have an effect on the instantiation of an object or a relation between objects. Hence, the instantiation constructors for categories and relations have to be specified either together with a probability value if the uncertainty value arises from a priori or statistical probabilities or together with a belief value if the uncertainty value arises from expert-defined or agent-based beliefs. These probabilities and beliefs in ontological instantiations are illustrated in table 5.2.

In order to integrate these uncertainty types, the ontology constructors for type and property thus have to be extended by probabilities and beliefs. The interpretation of both uncertainty types applied to the constructors are based on standard probability and evidence theory [Dubois and Prade, 2001]. The resulting modified constructors and their interpretation are shown in table 5.3. As both uncertainty types assign a value between 0 and 1 to statements, probabilities and beliefs are illustrated together by the function m, that assigns the instances and instance relations their probability or belief values.

A spatial application that requires and applies belief values for its ontological instantiation is presented in section 6.3. In this application, agent-based instantiations of categories and relations are affected by uncertainties that are defined by belief values [Schill et al., 2009]. This belief is defined by using the Dempster-Shafer theory of evidence [Shafer, 1976], i.e., the semantics of a belief of an agent in a specific instantiation or relation is given by $m(A)$ in Dempster Shafer's theory, and Dempster's rule of combination supports reasoning with beliefs. In particular, the combination of all

Table 5.3: **Extended Uncertainty Instance Constructors within Modules.** An extension of ontology instance constructors with probability and belief within ontological specifications.

Extended Syntax	Extended Semantics	Interpretation
$m(a : A) = n$	$a \in A^{\mathfrak{J}} \to [0,1]$	category instantiation with belief or probability value
$m(\langle a,b \rangle : R) = n$	$a,b \in A^{\mathfrak{J}} \wedge \langle a,b \rangle \in R^{\mathfrak{J}} \to [0,1]$	property instantiation with belief or probability value

beliefs about the evidence of the type of an instance, e.g., $m(x : Refrigerator)$ and $m(x : Freezer)$, does not have to sum up to 1 (in contrast, for instance, to Bayes probabilities). Concrete values within the specification are provided by expert knowledge. If the instantiation of a category is uncertain because of ambiguous data from an object recognition process, the concrete uncertainty value from the result of this recognition process can be used. If such uncertainty values are unavailable, they can be approximated by averaged probabilities over category restrictions, e.g., if a category defines n relations and a relations of its instance are verified, a belief of $\frac{a}{n}$ is assumed.

5.2.4 Uncertainties across Ontology Modules

In contrast to individual specifications of an ontology module, relations across ontologies for combining ontology modules are more likely to be affected by uncertainties. This is also reflected by the high number of approaches for uncertain mappings between ontologies in table 5.1. The combination of different ontology modules can be influenced by uncertainties in different ways: if ontologies are combined that comply with different perspectives, their combinations cannot be strictly defined due to their specification distinctions; if ontologies are aligned by using an automatic extraction based on statistical results from extensional mappings, their combinations may only define possible alignments to some degree; if ontologies are connected that differ in purpose, aims, and scope, their combinations might only indicate similarities between categories and relations without clear correspondences.

Hence, ontological constructors that allow combinations of, or relations between, different ontology modules have to provide an integration of uncertainty types. As introduced in section 5.1, several techniques are available for ontology combinations. Their technical implementation can often be reduced to OWL constructors for importing ontologies. Categories and relations of imported ontologies can then be related by using equivalence constructions on categories and relations. More generally, parts of one ontology are related to parts of another. For clarity, we use $\equiv^*$ as a placeholder for the combination constructor between two different modules. In OWL, the equivalentClass and equivalentProperty constructors are used accordingly.

Table 5.4: Uncertainties affecting Ontology Constructors across Modules. DL constructors that are affected by uncertainties across ontological specifications.

Syntax	Semantics	Uncertainty Type	Example
$A_1 \equiv^* A_2$	$A_1^{\mathcal{I}} \leftrightarrow A_2^{\mathcal{I}}$	probability	mod1:SpatialAction and mod2:Motion are related with a probability of .8
		belief	The belief (of an agent) that mod1:Window is equivalent with mod2:Window is .9
		similarity	mod1:Region closely resembles mod2:Area with a similarity value of 2
$R_1 \equiv^* R_2$	$R_1^{\mathcal{I}} \leftrightarrow R_2^{\mathcal{I}}$	probability	mod1:hasPart and mod2:contains are related with a probability of .9
		belief	The belief (of an agent) that mod1:partOf is equivalent with mod2:locatedIn is .7
		similarity	mod1:above resembles mod2:over with a similarity value of 8
$a_1 \equiv^* a_2$	$a_1^{\mathcal{I}} \leftrightarrow a_2^{\mathcal{I}}$	probability	mod1:user1 and mod2:Bob are related with a probability of 1
		belief	The belief (of an agent) that mod1:park1 is equivalent with mod2:recreationalArea1 is .2
		similarity	mod1:region1 resembles mod2:France with a similarity value of 1

The equivalence constructor indicates that parts from different ontologies are related with each other, which could be either in terms of identity, similarity, or some other connection. Depending on the interpretation of the combination relations, combinations thus need to be extended by probabilities or likeliness of the combination relation or by using similarities between parts. As a consequence, the $\equiv^*$ operation is affected by uncertainties that can be defined by probabilities, beliefs, or similarity values, illustrated in table 5.4.

In detail, inter-ontology relations between categories, relations, or instances are assigned uncertainty values either in terms of a probability, belief, or similarity type. Probabilities can be used, for instance, if combinations between ontology modules are based on statistical or external data, i.e., the likelihood of related categories, relations, or instances is implied with a certain probability value. Probabilities are an adequate representation primarily between modules that share the same perspective, as these may share similar meanings when categories across modules are combined or related. Beliefs are applicable particularly when agents or systems infer combinations between ontology modules. In particular, ontology-based agent communication is designated for using belief values to relate ontology parts (categories and relations) from different modules, regardless of the ontological perspectives used. Similarities can be used if relations between modules from different perspectives are described, as these modules are highly disjoint in their meanings and specifications. For instance, a connection between modules across perspectives can provide a functional combination, i.e., it reflects how similar two categories, relations, or instances are. As already discussed in section 5.2.2.1, fuzziness is not relevant in or across ontological representations. The overall constructor extensions are shown in table 5.5.

Table 5.5: Extended Uncertainty Constructors across Modules. An extension of ontology constructors with probability, belief, and similarity across ontological specifications.

Extended Syntax	Extended Semantics	Interpretation
$m(A_1 \equiv A_2) = n$	$(A_1^{\mathcal{I}} \leftrightarrow A_2^{\mathcal{I}}) \to [0,1]$	related categories across modules with probability or belief value
$sim(A_1 \equiv A_2) = n$	$(A_1^{\mathcal{I}} \leftrightarrow A_2^{\mathcal{I}}) \to n,\ n \geq 0$	related categories across modules with similarity value
(analogous to relations and instantiations)		

Concrete values for the uncertainties between ontology parts from different modules can be defined by ontology developers, expert users, agents that provide inter-ontology relations themselves, or by automatically detected inter-ontology relations. For example, a spatial system that uses a spatial domain ontology together with an ontology for qualitative spatial relations can define the likelihood of related categories, relations, and instances between the two ontologies. Given particular categories, such as Column, locatedIn, Room, the spatial system can then connect them to categories from the qualitative ontology, such as a specific Region that is a proper part of another Region, by assigning specific uncertainty values to each connection. In terms of the inference process, the interpretations of the uncertainty types applied to the equivalence constructors are based on standard probability, evidence, and similarity theory [Dubois and Prade, 2001; Sheremet et al., 2007]. Reasoning with evidence is discussed in more detail in section 6.3 for applying the Dempster-Shafer theory together with ontology modules.

A spatial use case that relies on the specification of uncertainties across ontology modules is presented in section 6.4. Here, the connection between two modules is affected by uncertainties, because the modules describe the spatial domain from different spatial perspectives. One module provides information from a qualitative spatial perspective, whereas the other module provides information from a multimodal spatial perspective. Both representations model spatial relations between entities, of which some resemble each other more closely than others. The uncertainties in the combination of both modules are thus represented by similarity values as their spatial perspectives highly differ [Hois and Kutz, 2008a]. Concrete values can be gained from statistical corpus data or expert knowledge.

5.2.5 Summary of Uncertainties in and across Ontology Modules

The different approaches for integrating uncertainty types in and across ontology modules provide methods for taking into account uncertainty aspects as far as appropriate with regard to the semantics of ontology constructors and their applicability. The uncertainty integrations, however, not only allow the representation of different types of uncertainties but also limit possible occurrences and effects of uncertainties. An ontological representation of a certain reality can accordingly be analyzed with respect to these permissible uncertainties that can be specified whenever adequate.

For this purpose, ontology constructors were extended with uncertainty types for the required values and aspects of uncertainty. In particular, probabilities, beliefs, and similarities have been integrated in the ontological specification language in order to model uncertain instantiations and inter-ontology module combinations. Their interpretation can be used according to standard probability, belief, and similarity theories, and the constrained use of uncertainties reduces complexity with regard to their possible (mis-)use in ontological specifications.

The uncertainty representations for ontological specifications are used in different spatial applications, as presented in the next chapter. For describing their domains, the applications combine ontology modules with different types of uncertainties. We show that the uncertainty constructions provided are sufficient for spatial application requirements.

5.3 Chapter Summary

In this chapter, two main techniques for extending ontology modules have been introduced, namely methods for combining different modules and methods for representing uncertainties in and across modules. The combination techniques not only support the various ways of combining ontological information from different modules with each other; they also reflect the possible interpretations of combining modules, i.e., either by integrating ontological information through extensions and refinements, by aligning ontological information through matchings, or by linking ontological information through connections. The uncertainty techniques allow the representation of different uncertainty types in and across ontology modules; they support probabilities, beliefs, or similarities for module instantiations and inter-ontology relations.

Ontology combinations as well as uncertainty aspects are both relevant when putting modules into application contexts. Combinations of modules are required for representing information about the application domain, and uncertainties occur when dealing with the different causes of uncertain knowledge within an application. Both, ontology combinations and uncertainty integrations, have been applied in different spatial application scenarios, which are presented in the next chapter.

6

Application and Evaluation

The modular ontologies for spatial information presented have primarily been developed to support task-specific requirements in spatial assistive systems and to provide a knowledge layer for spatial applications for making their ontological categories and relations transparent and thus (re-)usable. The following applications show how the spatial ontologies presented in chapter 4 and their extensions for combinations and uncertainties presented in chapter 5 can be used in different scenarios and as a support for application-specific purposes.

In particular, three different applications are presented, which show how spatial ontology modules are selected, used, and evaluated. The first scenario examines architectural design and assisted living and describes how the different module combinations can support their space-related tasks and requirements. The second scenario exemplifies the use of spatial modules to provide ontological information about space for visual scene recognition under uncertainty. The third application demonstrates the interpretation of spatial language in terms of spatial qualitative representations by using spatial module combinations.

The next section starts with aspects for selecting and applying the spatial ontology modules in spatial systems. It identifies the criteria each application has to define and analyze with regard to specific ontological requirements. It also examines how applications have to distinguish and select particular spatial perspectives and modules.

6.1   Selecting and Evaluating Modules

Similar to the development process of ontology modules, a requirement catalog needs to be specified by each application, i.e., each system describes its purpose, requirements, and task examples. Furthermore, it has to specify which domain aspects the ontological specification should represent, and which functions the ontological specification should provide and support in terms of the application. To this end, the spatial system is also assisted by classifying its requirements with regard to the spatial perspectives as qualitative/quantitative, abstract, domain-specific, and multimodal.

Hence, the three main application scenarios that are presented below first introduce their respective tasks, aims, application examples, and ontology module requirements. In detail, they specify which spatial ontology modules are required, how the modules should satisfy application-specific tasks, and whether ontology extensions in terms of module combinations or uncertainties are necessary. The applications also identify how the ontology modules can be applied and how their applicability and adequacy can be evaluated in the application contexts. The evaluations also indicate the modules' clearness, practicability, and reusability; however, the particular evaluation process has to be determined individually for each application.

In section 4.2, the ontology modules were already evaluated independently from applications; for instance, their consistency, satisfiability, reuse of existing ontological specifications or external sources, or documentation and availability were analyzed. For evaluating the modules' performance in the application context, their usability or cognitive adequacy is evaluated by using different methods. How well an ontology module represents a given domain can, for instance, be analyzed by measuring user agreement or system-internal domain agreement with the ontological domain model, which can also be provided by using representation standards or external sources in the ontological specification. It can also be measured whether or how well an ontological module supports the application's requirements or a specific use case. Here, pre-defined scenarios determine which information should be available from the ontology modules or which answers they should give to specific ontological queries from the system. The general performance of an ontological specification can also be measured by analyzing the overall system's performance with and without using ontological modules. Also a golden standard can be defined that specifies designated input and output data for ontological classifications or querying.

Therefore, the evaluation methods and the validity of their results are discussed for each application scenario individually. In addition, each application analyzes how clearly spatial perspectives can be distinguished and applied, in particular, as the perspectives are also used as a selection criteria to decide on the modules that are applied in a system.

6.2 Architectural Design and Assisted Living

The design of architectural environments and the development of assisted living have recently become AI topics that are built and extended with AI-related techniques [Ramos, 2007]. For instance, architectural design systems can be enabled to model indoor environments for analyzing navigational usability [Stahl and Schwartz, 2010], for inferring expected behavior or design consequences [Schultz and Bhatt, 2010], or for analyzing structural patterns in architectural designs and their similarities [Jupp and Gero, 2006]; assisted living environments can be implemented with semantic representations for supporting activity recognition in smart homes [Liao et al., 2010], for monitoring and controlling assisted living [Krieg-Brückner et al., 2009], or for automatically detecting user actions [Chua et al., 2009].

Ramos [2007] summarizes that several techniques from AI, such as machine learning, planning, natural language interaction, knowledge representation, or computer vision, can be used to achieve such goals. However, "little attention has been paid to Knowledge Representation in most of the Ambient Intelligence projects" [Ramos, 2007], and in many cases high-level information about objects or actions related to the environment is less based on formal ontology design principles or available resources but rather based on a given system and is thus often not generalizable. For the purposes of exploiting knowledge representation and achieving generalization, we analyze ways to develop and integrate ontological representations for architectural design and assisted living. As architectural design and assisted living applications are related to their spatial environments, formal representation and reasoning for spatial and domain-specific information is necessary to model the application-specific requirements and to enable their intended form and function.

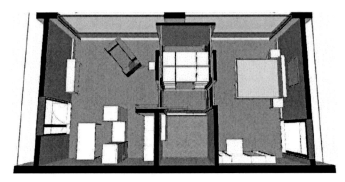

Figure 6.1: Assisted Living Environment. The BAALL floor plan, which shows the layout of the apartment. (Figure adapted from http://www.baall.net, visited on March 10, 2009)

Figure 6.1 shows an example of an assisted living environment, namely the Bremen Ambient Assisted Living Lab (BAALL), which has been designed for the elderly and people with physical or cognitive impairments [Krieg-Brückner et al., 2009]. The BAALL provides building automation, in particular control of lighting, air conditioning, appliances, doors, access restriction, and user-based profiles. The apartment is also equipped with intelligent furniture that can be automatically customized to the users. For example, cupboards in the kitchen automatically change their position to adjust to wheelchair users. As the BAALL has been developed particularly for the elderly, it also provides health-critical components, e.g., bio-sensors that can measure the heart rate, temperature, and general body activities. Finally, the apartment is intended to provide an adaptive assistance system that learns and creates user profiles and that aims at being comfortable and non-intrusive.

The example illustrates the two main topics in this section: the architectural design of indoor environments and the automation and monitoring of assisted environments. Both topics are closely related but focus on different aspects of a spatial environment. Architectural design addresses constructional characteristics of an environment. These can range from basic structural form-related to abstract functional or behavioral characteristics. For example, a basic design characteristic specifies that a room has a window and a door and a complex design characteristic specifies that a lobby is easily accessible and has a waiting area and a reception. The first specification can be modeled by directly using respective domain-specific ontology categories, the second specification needs to interpret the functionality of 'accessibility' in terms of the domain-specific categories. Assisted living, on the other hand, addresses functionalities that are intended to support human users in the environment. These functionalities can provide automatic monitoring and modification of basic environmental aspects, or they can provide high-level or even learned behavior. For example, an assisted living environment can automatically regulate the air conditioning according to user-specific preferences, or a smart work environment can automatically restrict or grant access for user groups to parts of a building. Although the assistance functionalities provide high-level behavior or adaptation of the spatial environment, they are based on environmental objects and architectural facilities that support the intended functionalities. For example, automatic air conditioning can be modeled with ontological categories that provide information about the spatial environment, such as users and their positions, temperature and heating, and position of air conditioners. Hence, both architectural design and assisted living specify and restrict similar domain-specific categories only with a different focus.

In the application scenarios presented below, the characteristics of architectural design and assisted living are formulated with spatial ontology modules. They model spatial and domain-specific information and their requirements to satisfy certain designs or assistance functionalities, which are often left implicit:

> 'design knowledge' (...) is unstructured and entirely devoid of theory. Instead it is implicit, practical and predictive. It enables us to recognize situations and tells us how to behave in them. It is largely implicit rather than explicit knowledge, since we seldom externalize it, rarely speaking of it and hardly ever writing it down. In fact we are usually not even aware we have this knowledge. [Lawson, 2001, p. 199]

The spatial perspectives (introduced in section 3.2) that can contribute to the specification of architectural design and assisted living are the qualitative and quantitative spatial perspective, the abstract spatial perspective, and the domain-specific spatial perspective. In the context of architectural design, constructional information can be specified by using qualitative or quantitative categorizations (e.g., for rooms, walls, windows, doors) and high-level functionalities can be specified by using abstract and domain-specific categorizations (e.g., for lobbies, accessibilities, waiting areas, receptions). In the context of assisted living, similar types of information are needed for architectural aspects, and additionally, abstract and domain-specific categorizations provide information for specific assistance behavior, e.g., to specify requirements for lighting, temperatures, or access restrictions.

The spatial perspectives lead to the following modules required for modeling architectural design and assisted living:

Quantitative and Qualitative Perspective. The modeling of constructional and structural information is based on quantitative data of floor plans and qualitative data of relational information in floor plans. Such ontology modules can be based on existing architectural data formats, e.g., the IFC [Liebich et al., 2010] as used in the *ACO* module (section 4.2.1.2), and on qualitative spatial representations, e.g., the RCC [Cohn et al., 1997a] as used in the *RBO* module (section 4.2.1.1). Both modules are used to model architectural design and assisted living.

Abstract and Domain-Specific Perspective. The modeling of conceptual information in architectural and assisted living environments is based on abstract and domain-specific categorizations that reflect the environmental objects and functions. Terminological information related to architectural floor plans and their structural functionalities is specified by modules complying with the domain-specific spatial perspective, such as the *ADO* module (section 4.2.3.2). The environmental objects are specified by employing a domain-specific perspective, which can refine an abstract perspective, such as the *PEO* module reusing DOLCE (section 4.2.3.1) to model physical entities. In assisted living environments, environmental behavior and functionalities are modeled by domain-specific spatial perspectives, such as the specification of home automation in the *HAO* module (section 4.2.3.3), which is based on abstract categorizations of behavior and actions, specified in the *SAO* module (section 4.2.2.1).

The two scenarios that we describe and evaluate in the following present the application of ontological modules in architectural design and assisted living for specifying and constraining intended structures and functionalities. For architectural design, we use the ontological combination of quantitative and qualitative as well as qualitative and domain-specific spatial perspectives. For assisted living, we use the specification of environmental requirements in terms of abstract and domain-specific perspectives.

6.2.1 Modeling Architectural Design

Architectural design tools allow designers to develop models of spatial structures at different levels of granularity. They typically support low-fidelity planar layouts as well as complex high-resolution 3-dimensional models to represent building plans. CAD tools, for instance, allow the design of floor plans with regard to spatial elements representing doors, windows, rooms, etc. However, these elements are based on primitive geometric entities and their terminological characteristics remain undefined [Bhatt and Dylla, 2009]. As a consequence, such architectural tools lack the ability to exploit the conceptual expertise of designers (comprising architects, engineers, town planners, urban designers, interior designers, or landscape architects [Lawson, 2001]), which is a limitation considering the new generation of building automation systems and smart environments [Hois et al., 2009b].

For example, the architectural tools are not capable of modeling the requirement that a motion sensor should be placed in a way to entirely cover a door with its range space, e.g., for monitoring reasons. Furthermore, conventional design expertise is often driven by experience and intuition, and it is concerned with spatial and structural aspects of the design rather than its functional characterization. Our ontological modeling of architectural design aims at filling these gaps by specifying and combining qualitative, quantitative, and domain-specific spatial information [Hois et al., 2009a]. How architectural information can be analyzed and how design functionalities can be interpreted in terms of these perspectives is illustrated by the following architectural requirement available in a design guide for courthouses:

> **Witness Box.** *Witnesses must be able to see and hear, and be seen and heard by, all court participants as close to full face as possible. The witness box must accommodate one witness and an interpreter (...). Witnesses in the box receive, examine, and return exhibits. The interpreter must be seated next to or slightly behind the witness and between the witness and the judge; however, the witness must remain the primary focus. A separate*

interpreter station, accommodating two staff persons, might be required in
some locations. [US GSA, 2007, p. 4-13]

These requirements that define spatial constraints and relations for a witness box in a courtroom demonstrate the complexity and abstract functionality prevalent in architectural design [Bhatt et al., 2012]. As pointed out above, CAD tools are neither able to visualize such requirements in a floor plan nor able to analyze whether a floor plan, e.g., the courtroom example in figure 6.2, satisfies these requirements, as the tools lack the ability to represent non-quantitative spatial elements and functionalities. Hence, the purpose of our ontology modules for architectural design is to distinguish the different types of spatial information in architectural design according to their spatial perspectives, to model and combine them by specifying respective ontology modules, and to formulate architectural functions and requirements by using the combined ontology modules.[1]

To achieve a complete ontological specification for architectural design, ontology modules with the following spatial perspectives have to be instantiated and connected with each other: a module complying with the quantitative spatial perspective that provides constructional information, which is directly available by the floor plan, namely the *ACO* module; a module complying with the qualitative spatial perspective that provides spatial qualitative information about the elements in the floor plan or the elements from abstract design requirements, such as the *RBO* module for region-based relations; modules complying with the domain-specific spatial perspective that provide semantics for the elements of the floor plan and their structural characteristics and functionalities, namely the *PEO* and *ADO* modules.

The combination of the individual modules results in the *EConnEntities* module, which is now extended to provide categorizations specifically for describing the requirements and constraints of architectural design, i.e., the extended module is an architectural requirements module. For example, the requirement that "the witness box must accommodate one witness and an interpreter" can be modeled by formulating specific restrictions over categories from different modules. Witness box, witness,

---

[1]Note that the interpretation of functions here refers only to those aspects that emanate directly from the structural form of architectural design, i.e., the functions are identifiable directly by spatial, physical, and conceptual constraints. Functions related to social, cultural, aesthetic, or economic constraints are beyond the spatial scope.

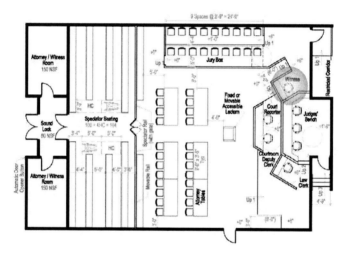

Figure 6.2: Courtroom Design Example. The figure shows a floor plan of a courtroom, in which the highlighted area indicates the witness box. The witness box follows the courthouse design guidelines, however, not all requirements are directly available from the spatial layout of the design. These need to be derived conceptually. (Figure adapted from [US GSA, 2007, p. 4-18])

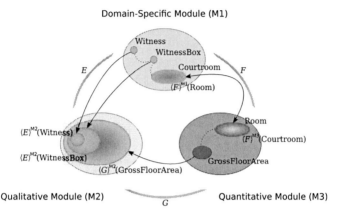

Figure 6.3: Three-Dimensional Module Combination. The three modules for quantitative (M3), qualitative (M2), and domain-specific (M1) spatial information model the different perspectives on architectural design, and their combinations allow the specification of design-specific requirements. Only the combination of the three modules can model, and analyze the satisfiability of, the requirement that witnesses be positioned inside witness boxes. The $\mathcal{E}$-connections E, F, and G provide the combination of categories from different modules, and together they result in an architectural design requirements module. The indices M1, M2, and M3 of instantiated $\mathcal{E}$-connections indicate in which module the $\mathcal{E}$-connections are interpreted.

and interpreter are categories from the domain-specific module. These categories are connected with their respective geometrical representations in the quantitative module. The respective regions in the qualitative module that correspond to these geometrical representations then need to satisfy certain requirements, namely that the regions of the witness and interpreter be a proper part of the region of the witness box.

As a result, the specifications for quantitative, qualitative, and domain-specific spatial architectural information are kept separate in the ontological modeling, and connecting these representations provides the specification of functional requirements. This not only reflects the different types of spatial information but also allows the specification of architectural functions and requirements and their relationship in an appropriate way. In particular, these are specified by using ontological constraints over categories from different modules (using module combination methods introduced in section 5.1), illustrated in figure 6.3.

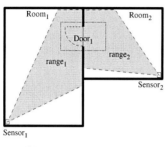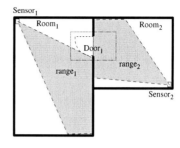

(a) Requirement Consistency   (b) Requirement Inconsistency

Figure 6.4: Architectural Design Scenario. A two-room scenario with the requirement that the door must be monitored by motion sensors. The entire space of the door thus needs to be completely covered by some (not necessarily one) motion sensor ranges, i.e., the door space region is a proper part of the range space regions. (Figure taken from Bhatt et al. [2009])

We will demonstrate the technical ontological specification of architectural requirements and their satisfiability in the following architectural design example: *a door should be monitored by some motion sensor devices in a way that there are no blind (hidden) areas of the door's doorway, i.e., it has to be ensured that motion sensors properly perform a 'handshake' when monitoring persons changing rooms through the door* [Bhatt et al., 2009]. Figure 6.4(a) illustrates a floor plan that satisfies this requirement, as the entire space of the door is covered by two motion sensors from different rooms. Figure 6.4(b) illustrates a floor plan, in which the sensor ranges do not entirely cover the space of the door, and consequently the requirement is not satisfied.

The different categories for Room, Door, MovementSensor, and RangeSpace are specified by the *ADO* module, shown in listing 6.1. Instances of the first three categories necessarily have connections to their quantitative counterparts in the *ACO* module, which are specified by the eConnComposedOf relation from the *EConnArchitecture* module.

Listing 6.1: Specification of specific architectural requirements for motion sensors

Annotations:
> rdfs:comment ' Parts of the ADO module that specify
> the categories for rooms, doors, and motion sensors
> and their ranges. ' @en

Class: Room
> **Annotations:**
> > rdfs:comment ' The most basic element that is part of a building
> > is defined by a room. Its counterpart region in the RBO module
> > defines its region−based constraints with regard to the building,
> > in which the room is located. ' @en
>
> **SubClassOf:**
> > BuildingPart,
> > consistsOf **some** Door,
> > consistsOf **some** Wall,
> > RBO:connectsWith **some** (Corridor **or** Room),
> > RBO:properPartOf **exactly** 1 Level

Class: Door
> **SubClassOf:**
> > BuildingConstructionPart

Class: MovementSensor
> **SubClassOf:**
> > SensoringDevice,
> > rdfs:comment ' The following definition is inferred from
> > the supercategory: The sensor is related to its range space
> > of its sensing (functional) area. ' @en
> > hasFunctionalStructure **exactly** 1 RangeSpace

Class: RangeSpace
> **SubClassOf:**
> > FunctionalStructure,
> > rdfs:comment ' The following definition is inferred from
> > the supercategory: A range space can only exist for one
> > specific building structure, e.g., a sensing device. ' @en
> > isFunctionalTypeOf **exactly** 1 BuildingStructure

An instantiation of a floor plan consists of instances in the *ADO* and *ACO* modules together with connections to instances in the *RBO* module by using relations from the *EConnArchitecture* module. This way it can be analyzed if the floor plan satisfies the 'handshake' requirement. In figure 6.4(a), the space of $Door_1$ is a proper part of the union of the range spaces of $Sensor_1$ and $Sensor_2$. It therefore satisfies the requirements and is proven to be consistent. We technically verify this with the RacerPro reasoner [Racer Systems, 2007], which provides S-Box reasoning for reasoning over RCC relations that were used for the architectural regions [Bhatt et al., 2009]. For instance, we can analyze that no instance exists ('NIL') that is the region of a door without being a non-tangential proper part of range spaces of motion sensors.[2] The query in RacerPro infers this result as follows:

```
? (retrieve (?*X ?*Y) (and   (?X DoorFunctionalSpace))
                              (?Y MotionSensorRangeSpace)
                              not (?*X ?*Y :ntpp)))
> NIL
```

Here, DoorFunctionalSpace specifies the region of the space around the door (illustrated in figure 6.4(a)), and MotionSensorRangeSpace specifies the region of the range space of the movement sensors. The same request with the example in figure 6.4(b) infers that $Door_1$ is not a non-tangential proper part of sensor ranges, and the design is accordingly inconsistent with respect to the 'handshake' requirement.

In summary, concrete architectural designs can be specified by reusing the combinations primarily of the ADO and ACO modules. The different perspectives contribute to the overall representation of architectural elements and their characteristics. Specific designs can be analyzed whether they satisfy certain requirements by using reasoning over one or several modules. The example has shown how qualitative region-based reasoning was used to test the requirement consistency of an architectural design. In addition, standard OWL reasoning over TBox or ABox provides the analysis of ontological consistency and satisfiability of architectural designs. Although design requirements vary between specific applications, e.g., for design requirements for Chinese architecture [Wei et al., 2010], the spatial modules can be reused and extended to provide the required forms and functionalities for respective designs.

---

[2]The racer specification of the example is available at `http://www.informatik.uni-bremen.de/~joana/ontology/modSpace/reasoningExample.racer`

6.2.2 Regulating Assisted Living

Assisted living environments aim at providing building automation, such as control of lighting, air conditioning, appliances, doors, and access restriction [Mann, 2005]. These functionalities, also called *ambient intelligence*, have a direct effect on the environment and the entities contained therein, and they typically cause a change of these entities or their properties. Ontology modules that aim at specifying these environmental entities and functionalities for home automation therefore have to take into account several types of spatial information. Information about architectural elements of the home environment, such as walls, doors, or windows, provides the most basic type of architectural data necessary to describe the spatial environment of an assisted living scenario. Information about the entities that a home environment is equipped with, such as appliances, devices, pieces of furniture, or sensors, is necessary for describing what is contained in the indoor environment and how the environment can be changed and affected. How these environmental entities are positioned in or related to the architectural environment can be specified by connections between both types of information. In addition, information about users and their actions is relevant to describe their influences on the environment. And finally, information about the actual building automation and its functionality is necessary to specify the entire assisted living scenario.

Automatic control of different electronic devices can be supported by their ontological specifications. For instance, home automation can be specified by constraints over environmental entities for lighting, temperature, or access. At certain brightness or temperature values, which are ontologically specified by environmental entities, sun-blinds may open or close, which is ontologically specified by spatial actions. Heating of air and water measured with sensing devices, which are ontologically specified by environmental entities, can be regulated, which is ontologically specified by home automation. Automatic locking of front doors, which are ontologically specified by architectural features, can be defined, for instance, if the house owner is at home, which is ontologically specified by physical roles and spatial locations. Visual sensors can be used to determine particular users to control access, which is ontologically specified by home automation. [Hois, 2010c]

As introduced in section 4.2.3.3, we developed the *HAO* module specifically for assisted living, which extends a combination of several spatial ontology modules to provide the required environmental information and to specify particular home automation functionalities. Here, we discuss how the combined ontology modules contribute to the description of the assisted living scenario, such as the one introduced in Krieg-Brückner and Shi [2009].

The *ACO* module specifies the architectural entities that are required for basic spatial information of the assisted home environment, i.e., the module indicates how the environment is designed and which constructional elements the environment contains. As introduced before, categories from the *ACO* module reflect information available in construction plans. A Door, for instance, is defined as an area that is attached to wall, floor, or ceiling areas by using region-based relations from the *RBO* module. It is defined by its spatial extent (length, height, and width) and an opening radius. Possible occurrences of doors are constrained by the *HAO* module based on their connections to other areas. For home automation, this information can be used to monitor the status of constructional elements, e.g., whether doors and windows are open or locked (causing effects on heating or access functionalities).

The *ADO* module specifies structural characteristics of the home environment, i.e., it provides the functions available from the architectural design. For example, information about electrical devices, such as sensors, is specified in this module. The combination of the *ADO* and *ACO* modules results in the *EConnArchitecture* module, which provides the types and locations of sensor devices the environment is equipped with. Here, sensors for air conditioning and their effects on temperature regulations can be adjusted depending on the status of doors and windows if necessary.

The *PEO* module defines physical entities in indoor environments. These entities also occur in assisted living environments, for example, specific subtypes of appliance, furniture, or user. Accordingly, the *PEO* module can support the specification of objects particularly for assisted living. For instance, information messages that an InformationBoard displays to users can be configured as required. Such information types are available in the *PEO* module.

The combination of these different ontological modules can also prevent ontologically contradicting definitions. For instance, doors in the *ADO* module are defined as (entrance) connections between rooms, that are consequently externally connected to the adjacent rooms by using the respective qualitative relation from the *RBO* module. Doors in the *PEO* module, however, are defined on the basis of their physical properties, i.e., size, weight, material, or color. While the structural door needs to be connected with at least one room, the physical door does not. While the physical door can be made of a certain material, the structural door cannot. Hence, the definition of 'Door' in one module contradicts and does not match with the definition of 'Door' in the other module, as both categories cannot be subsumed by (or aligned with) each other. Consequently, an instance of 'Door' in one module would not be an instance of 'Door' in the other module. By using different modules the different definitions can be separated, even though they describe the same door in reality, and the contradiction is avoided. Two instances for the same door entity from two thematically different modules can be linked with each other by a connection between the modules, as specified in the *EConnEntities* module.

The *SAO* module defines possible actions that can be performed in indoor environments. Hence, the module provides general information about actions of users, events, and states of entities. For example, an instance of Door from the *PEO* module can be open, closed, or locked. This status may change because of an event or a user interaction; automatic access restrictions lock and unlock doors automatically, while users leave doors open, closed, locked, or unlocked. These requirements are specified in the *HAO* module, which extends the *EConnEntities* module and the *SAO* module, in order to provide particular functionalities of an assisted living environment.[3] In contrast to the other ontologies, particular definitions, e.g., user-based access restrictions, depend on the specific building they are applied to, and are thus application-specific. Hence, modules similar to the *HAO* module need to be re-defined for different environments and functionalities.[4]

For an assisted living application, the modules are distinguished by means of their spatial perspective and their thematically different semantic description of the assisted living environment. They reflect quantitative, qualitative, abstract, and domain-specific perspectives on space, as illustrated in figure 6.5. One of the example scenarios

---

[3]Although it is possible to use rules to specify the home automation requirements, these can be reduced to role inclusions, which is further discussed in section 7.2.

[4]In order to interact with the assisted living environment, natural language may be used for human-computer interaction. Categories and relations for spatial language are specified by the *GUM-Space* module and could be related to the *HAO* module, however, this is left for future work.

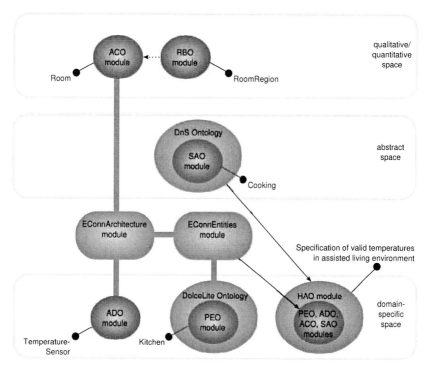

Figure 6.5: Modules and their Combinations for Home Automation. The figure shows the spatial ontology modules for specifying assisted living home automation. As an example, the individual categories that are required from the different modules to formulate valid temperature values in kitchens are shown.

for temperature regulation in assisted living environments shown in listing 6.2 (the specification is similar to the bathroom specification in listing 4.11, section 4.2.3.3) relies on the different categories from the modules: unless a cooking action (SAO) happens in (HAO) a room (ACO) that is a kitchen (PEO), the environment allows valid temperature values of temperature sensors (ADO) to be 16–20°C (HAO).

Listing 6.2: Specification of temperature values in kitchens

Class: PhysicalEntity:Kitchen

    rdfs:comment ' The HAO module specifies valid temperature values for instances of Kitchen in an assisted living environment. 'Abnormal' temperature values are only allowed if a Cooking action happens (and the instances are consistent with regard to the module specification). ' @en

    SubClassOf:
      (inv (happensIn) some SpatialAction:Cooking)
      or (EConnEntities:eConnRealizes **only** (PhysicalEntity:Kitchen
        and (inv (BuildingStructure:functionalType)
          only (BuildingStructure:Room
          and (RBO:inverseProperPartOf
            some (ArchitecturalConstruction:TemperatureSensor
            and (BuildingStructure:sensorValue
            some { ' 16C−20C ' ^^xsd:string })))))))

As a result, assisted living requirements or monitoring functionalities are modeled specifically by axiomatizing the categories from different modules and by using module combination relations. However, individual home environments need to reformulate their respective home automation characteristics and constraints. Some of such examples were presented in Hois et al. [2009a] and Normann et al. [2009].

In summary, for architectural design and assisted living especially, all ontological modules define the same environment by their different perspectives, however, all of them refer to entities of the same real assisted living environment. In particular, the modules are separated along their spatial perspectives on the domain and they keep apart different interpretations and meanings of categories and relations. For example, two modules may specify the same floor plan but one with a constructional and one with a design focus, which can even handle contradicting implications, as described above. A basis for communication and exchange is also provided among different user groups: the ways an architect, an interior designer, a construction worker, a painter, or a resident talk about walls differ as their perspectives differ, and these perspectives have been represented by distinct ontology modules and their combinations.

6.3 Visual Recognition

Visual recognition is a field of Artificial Intelligence that combines low-level image processing with high-level object and scene classification. Its aims are to recognize single objects in an input image or a video stream [Forsyth and Ponce, 2003] in order to classify object types and interpret image situations [Schill et al., 2009], track objects in different images or in a video stream [Bennett et al., 2008], provide question answering [Katz et al., 2004], allow object manipulation [Kjellström et al., 2008], support text to scene and scene to text translations [Gerber et al., 2002], or make event and action predictions [Yuen and Torralba, 2010]. To achieve these goals, visual recognition applies biologically-inspired image processing techniques that are motivated by human visual and cognitive perception. In particular, the field of *cognitive vision* pursues this idea:

> The term cognitive vision has been introduced to encapsulate an attempt to achieve more robust, resilient, and adaptable computer vision systems by endowing them with a cognitive faculty: the ability to learn, adapt, weigh alternative solutions, develop new strategies for analysis and interpretation, generalize to new contexts and application domains, and communicate with other systems, including humans. [Auer et al., 2005, p. 1]

When visual recognition systems integrate sensorimotor capabilities for image understanding, this is called *active vision*. Such systems use an additional motor control that can change location and orientation of their sensors to actively explore and analyze visual scenes [Bajcsy, 1988].

Figure 6.6 shows a visual recognition example. The figure shows image data that is represented by a 3D point cloud laser scan and it shows 2D photos of the related scene. Here, the task of the visual recognition process is the scanning of the scene, the detection of cloud clusters for the segmentation of objects, and the identification of objects based on the segmented bounding boxes [Hois et al., 2007]. The identified objects are learned to be classified as pre-defined object types, i.e., a training set of bounding boxes and their annotated labels, e.g., 'desk', 'bottle', 'book', are used to learn these object types. In this scenario, the object classification is particularly based on plane-based functionalities, namely the distance of objects to the floor, and dimension-based functionalities, namely the measurements of object classes [Wünstel, 2009]. These visual features classify object types with a probability value for the degree of object type membership of the objects. These results, however, are limited to the analysis of single objects and their low-level visual features and become less robust as the object domain increases.

To resolve this constraint, visual recognition combines visual feature analyses with high-level background knowledge of the visual data for improving classification results. Background knowledge provides information about visual objects that can occur in a scene, their relations to each other, and their implications for the type of the scene. As it "is evident that scene interpretation is a knowledge-intensive process which is decisively shaped by (...) common-sense knowledge and experiences" [Neumann and Möller, 2004], ontological representations can provide a structure and classification of the domain knowledge to support the visual recognition process [Maillot and Thonnat, 2008].

Neumann and Möller [2004], use the following aspects provided by background knowledge for high-level scene interpretation: the objects and occurrences that are involved in a scene; temporal and spatial relations between parts of a scene; qualitative and geometrical relations between objects in a scene; contextual information that is known about the scene; inferred facts that can be observed in a scene; and conceptual knowledge and experiences about the world, which is known in advance. In the use cases described in this section, we particularly focus on the first aspects, namely the objects that are detected in a scene and the scene type implied by the aggregation of these objects. The developed ontological specification uses abstract and domain-specific spatial perspectives in order to define scene-specific information. Caused by the infinity and variety of scenes as well as possible errors in the object recognition, we also address the integration of uncertainty aspects within the ontological representations. As a consequence, we analyze which uncertainty types affect the visual recognition and how they can be combined with the domain-specific ontological representation.

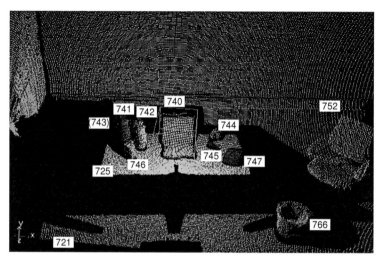

Figure 6.6: Visual Scene Example. The upper image shows the 3D representation of the lower images scanned with a laser sensor. The image is a result of a cognitive vision processing system [Wünstel, 2009] that segments and classifies objects in a scene [Hois et al., 2007]. The lower images show a part of an experimental office scene. The scene shows a desk with typical office-related objects on it. The position of the laser indicates the viewpoint toward the scene. (Figure taken from Hois et al. [2007])

Based on the separation of different spatial perspectives, the following ontology modules are applied or developed in the context of our visual recognition scenarios.

Abstract Perspective. Modules with high-level representations for conceptual objects support the description of the visual domain on an abstract level. They provide a general modeling of objects available in specific scenarios and abstract perspective modules can be reused by domain-specific modules for individual scenarios. We here reuse particularly the foundational DOLCE ontology (see section 2.6.3) to model domain-specific objects for visual recognition.

Domain-Specific Perspective. In order to specify a scenario-specific domain, the objects it contains, and their properties for visual recognition, we developed the *PEO* module, which reuses the DOLCE ontology. As introduced in section 4.2.3.1, the *PEO* module complies with the domain-specific perspective and provides information on physical entities. Even though these entities can also be understood as those objects that are visually perceivable, the *PEO* module does not specify them by their visual features, i.e., the *PEO* module neither complies with a multimodal perspective nor describes information on texture, colors, or shapes. In contrast, the physical entities are specified by their conceptual and spatial properties and their relations. The relation between objects and scenes, in which they occur, and their functionalities is specifically modeled in the *PEO* module. As described below, the module can further be reused or refined to adjust to specific vision recognition scenarios.

Note that an alternative approach to identifying scene types based on individual objects is to analyze the overall image as one category, which is called the *gist*, without segmenting and identifying single objects [Oliva and Torralba, 2001]. This way, scenes are classified as a whole on the basis of their visual features without classifying individual objects contained in these scenes. Nevertheless, high-level knowledge is required to model and classify the scene types. Also the (ontological) distinction between objects and scenes is lost in such approaches, although this information may be important for further processing, for instance, by a scene to text generation system.

The two application scenarios that are described in the following present visual recognition systems that apply ontological specifications for scene analysis. The first case presents how an ontological representation is developed and specified in order to support visual recognition and how uncertainty types can be combined with the ontological specification for a limited domain. The second case presents how statistical data is combined with an ontological representation for visual recognition and how the overall system performs its recognition tasks for a broader domain.

6.3.1 Image Recognition for Interior Rooms

The first application scenario integrates an ontological representation of domain-specific information with an active vision system that uses an active motor control of the position and orientation of its sensors in the analysis of visual data and exploration of spatial scenes [Schill et al., 2009]. In particular, the system identifies interior rooms by using an active exploration of the objects that are located in these rooms. It applies biologically-inspired techniques to classify low-level visual features and it combines these features with high-level learned features about specific objects and scenes [Schill et al., 2001]. Before we present the extension of this system with ontological information, we briefly summarize the active vision process.

The biological vision component aims at imitating human saccadic eye-movements to select and identify relevant visual features. This processing of visual input provides an essential component in the identification of images as performed by humans [Yarbus, 1967]. The visual recognition capabilities of the human eye are concentrated in a small region of the visual field, the *central fovea*, and rapid eye movements are performed to collect visual features from different relevant spots of an object. These spots, however, need to be extracted by means of their relevance for the identification of the object, i.e., primarily *salient* local features of an object are selected. This selection is driven by high-level cognitive processes, which rely on general world knowledge and on learned memory about specific objects.

These two components are implemented in an active vision system, presented in Schill et al. [2009]. The low-level sensory feature extraction uses neural operators that are optimized for natural scenes [Zetzsche and Krieger, 2001]. The high-level explorative selection of object regions that have the highest impact on the current internal belief about the identification of the object uses a belief-based reasoning strategy [Schill, 1997]. The combination of salient visual features and saccadic movements among these features represent the classification of objects, which is labeled with an a priori object type (e.g., 'flower' or 'chair'). For this object type representation the system uses an internal belief distribution over a hierarchically structured representation of object hypotheses. An object analysis cycle in the active vision system can be outlined as follows: first, salient features in an image are detected on the basis of low-level vision processing; after focusing on one of these features, the saccadic movement to the next feature is calculated based on its information gain; the combination of focused features and the motoric action among them result in sensorimotor features that update current belief distributions of object classifications; this cycle is repeated until a sufficient belief threshold for one of the object hypotheses is reached.

To analyze scenes that contain a variety of different objects that can be visually classified to a certain belief degree, we develop and apply a domain-specific spatial ontology module that reflects information about objects in scenes and their relations. Figure 6.7 illustrates the sequence of the visual recognition process, which integrates the ontological representation. For every object that is classified by its visual features, an ontological request determines possible scene type classifications. The respective results can be combined with an overall belief of the scene classification and further objects can be analyzed. The ontological module should therefore provide domain-specific information about interior rooms and their contained objects. It should also cope with the uncertainties caused by the belief-based detection of visually classified objects in the active vision system.

Hence, to support this vision process with an ontological representation the ontological module is required to specify interior rooms and the objects they contain. This information is taken from empirical data of indoor rooms and expert knowledge, which classifies the room's scene types to some degree of belief [Hois, 2006]. In section 4.2.3.1, we have presented a spatial domain-specific module that has been developed to be used in vision recognition systems, such as the one described above. This *PEO* module provides information about objects that are visually perceivable and it aims at supporting the recognition and classification of visual environments and the objects they contain. Furthermore it can be grounded in environmental data, namely the sensomotoric visual features for object classes given by the active vision system.

The main aim of enriching the visual system by an ontological representation is to identify complex scenes by the objects they contain. For example, the visual data shown

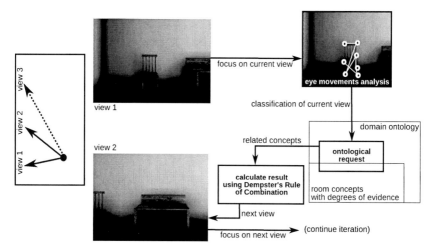

Figure 6.7: Active Vision integrating Ontological Representation. The left part of the images shows a birds eye view of the overall scene and the perspective from where view 1-3 are seen, which indicates an observer who focuses on different objects in a room. Every view contains an individual object that can be extracted and analyzed based on its visual senorimotor features, illustrated in the upper right part. The classification result is used to formulate the ontological request to acquire the scene types that are indicated by the classified objects. The resulting potential room scene types and their belief values are combined by using Dempsters Rule of Combination. An iterative analysis of further objects can improve the scene classification result. (Figure adapted from Schill et al. [2009])

in figure 6.7 is taken from a larger environment, which contains further objects, e.g., desks and bookshelves. The system identifies not only these single objects but also the overall scene, such as an office or workroom. These types of information about single objects and interior room concepts were specified in the *PEO* module. The module also provides an explicit distinction of objects versus scenes (cf. also Knauff [1997] for the perception of room concepts by humans) that is already used in the active vision system: As introduced in section 4.2.3.1, the *PEO* module classifies the objects that occur in the vision system as a subcategory of SpatialEntity, which is subsumed under DOLCE's physical-endurant and which has direct spatial properties and is independent from occurrences. Its subcategory FurnitureEntity, for instance, specifies subcategories for the objects chair, desk, and cabinet that occur in the interior room domain of the vision system. Furthermore, the *PEO* module specifies the scenes of the interior rooms perceived by the vision system as a subcategory of RoomType, which is subsumed under DOLCE's non-physical-endurant and determines the functional characteristics of a room. Its subcategory EquipmentRoom, for instance, specifies subcategories for the scene types office, laboratory, and copy room. As room types are identifiable by their functionality and this functionality depends on essential objects to support the function, recognized objects indicate the room scene in the vision system.

However, it may not be clearly specifiable whether a specific room scene necessarily has to contain certain objects, and also the visual system may not be able to visually recognize all relevant objects, e.g., because they are hidden behind other objects. This not only causes uncertainty affecting the domain-specific representation but also results in underspecified room types. For example, the ontological module can specify that an office 'can' contain a desk, a chair, and some other working-specific objects, but these objects do not necessarily need to be contained in an office. A visual detection of a desk and a chair will consequently not lead to any specific classification by the ontological modules. For a solution to this, we follow a different approach for specifying the domain-specific information together with its uncertainties: we 'out-source' the dependencies between room scenes and their objects to an ABox of the *PEO* module. This way, empirical data of a random set of room scenes is instantiated, which reflects the visual data by ontological instances, uncertainty values are instantiated together with visual data if necessary, and ontological queries for classifying scenes based on objects are performed by using ABox requests [Hois et al., 2006].

The ABox specifies objects that are classified by the visual recognition system as being located in a physical room. This physical room is related to potential room scene types together with a degree of uncertainty as the room can provide multiple functions. In many cases, room scenes can only be determined with a certain degree of belief precisely because of ambiguities in the relations among objects and scenes, i.e., a room scene is often compatible with more than just one room type. For example, if the vision system identifies a scene containing chairs and tables it may be confronted with a lecture hall, a seminar room, or a storeroom. A room scene is thus not restricted to identify only one room type. It can, for instance, provide the functionality of a lecture hall as well as a seminar room simultaneously by containing a sufficient number of chairs, tables, blackboards, and projection facilities. However, a room scene containing a refridgerator, a microwave, a sink, tables, and cabinets might not provide the functionality for a kitchen and a chemical laboratory simultaneously. Hence, the analysis of a room scene can only indicate the room type to a certain degree of belief.

This evidence is represented by using values of the Dempster-Shafer theory, described for ABoxes in section 5.2.3, because it reflects the belief of the active vision system with respect to its environment. A room scene is thus classified on the basis of empirical data from prototypic room samples, for which the room type is available [Hois et al., 2006]. Figure 6.8 illustrates the ontological instantiation of room scenes based on the *PEO* module categorization. Every classified object that has been detected by the vision system is instantiated as a subtype of the category SpatialEntity. These objects are related to their room scenes, in which they are contained, and these room scene instances are a subtype of the category PhysicalRoom. The functional type of the room scene is instantiated as a specific RoomType and related to the room scene. Room scenes can be related to several instances of RoomType to allow ambiguity and uncertainty, using the inverse relation of dolce:generically-dependent-on (cf. section 4.2.3.1 for more details). Also, instances of room type can specify a belief value with a data range from 0 to 1. This integrates degrees of belief into the ABox, i.e., the uncertainties are constrained to a particular domain sample and do not affect the *PEO* module specification. We instantiate the example for room1 and its relations in figure 6.8 as part of an ABox of the *PEO* module, which is shown in listing 6.3.

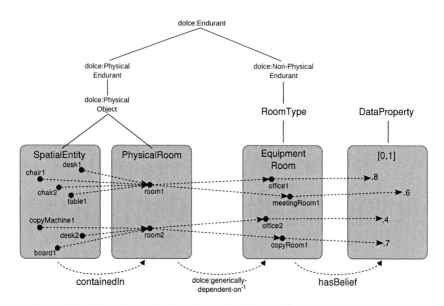

Figure 6.8: ABox Instantiations for Active Vision. The boxes contain the instances of categories; lines indicate subcategory relations; dashed arrows indicate relations between instances, i.e., ObjectProperties or DataProperties, and they point to the property range. Instances of the category SpatialEntity reflect the object of the visual recognition system. These objects are located in specific scenes, namely instances of the category PhysicalRoom. Instances of the category RoomType can only exist when related to a specific physical room, and room scenes can be inversely related with specific room scene types. To integrate the belief values of room types that are derived from objects contained in room scenes, every room type is related with one belief value that assigns a value 0–1, which models the belief value of the Dempster-Shafer theory.

Listing 6.3: Specification example of an ABox for vision recognition

Annotations:
> rdfs:comment ' The ABox imports the PEO module
> with the namespace PhysicalEntity, which imports
> DOLCE–Lite with the respective namespace.' @en

Individual: office1
> **Types:**
> > PhysicalEntity:Office
> **Facts:**
> > DOLCE–Lite:generically–dependent–on room1,
> > hasBelief .8

Individual: meetingRoom1
> **Types:**
> > PhysicalEntity:MeetingRoom
> **Facts:**
> > DOLCE–Lite:generically–dependent–on room1,
> > hasBelief .6

Individual: room1
> **Types:**
> > PhysicalEntity:PhysicalRoom
> **DifferentFrom:**
> > room2

Individual: desk1
> **Types:**
> > PhysicalEntity:Desk
> **Facts:**
> > PhysicalEntity:containedIn room1

Individual: chair1
> **Types:**
> > PhysicalEntity:Chair
> **Facts:**
> > PhysicalEntity:containedIn room1

Individual: chair2
> **Types:**
> > PhysicalEntity:Chair
> **Facts:**
> > PhysicalEntity:containedIn room1

Individual: table1
   Types:
      PhysicalEntity:Table
   Facts:
      PhysicalEntity:containedIn   room1

The belief value that is related by hasBelief is interpreted as $m(A)$ in the Dempster Shafer theory (cf. section 5.2.3). It is only specified in the ABox instantiation of the *PEO* module and not in the *PEO* module itself. Calculating the overall belief about a room scene applies Dempster's Rule of Combination using related belief values gained from the ABox query (cf. figure 6.7). In particular, singleton hypotheses are calculated for each RoomType (cf. Gordon and Shortliffe [1985] for more details on the singleton calculation). For example, if the results of two analyses for a room scene, e.g., from different points of view, are indicating that the room scene has the functionality of an Office, first with a belief value of 0.7 and second with a belief value of 0.9, the total belief for Office is 0.97 (the combination of both belief values).

This way, Dempster's Rule of Combination calculates current belief distributions iteratively, as illustrated in figure 6.7. The overall classification of a room scene type is split into the following subroutines: The first performs eye movements on the whole room scene, stopping at each fixation point (senorimotoric feature). The second extracts the object that is located at each point (e.g., view 1) and classifies its object type (SpatialEntity) by using eye movements based on a prior learning phase. Third, the object classification result is used for making a request to the *PEO* module's ABox; instances of room scenes that contain the classified object (after the first iteration together with previously classified objects) indicate potential room types with belief values. Their combinations are calculated and yield the room scene classification. Further iterations are performed (e.g., for view 2 and view 3) until an acceptable belief value is gained, e.g., a belief value above .9.

Interior room images that we used for room scene classifications by the active vision system are random samples that especially show university and office environments, taken from Hois [2006] and Hois et al. [2007]. Room scene functionalities are classified by experts with a certain degree of belief. The vision system was tested with these samples from 25 physical rooms that were instantiated as an ABox[5]. Even though the classification performance results in a belief of .97 on average after 5 iterations, which shows a high performance result, it is limited to the selection of domain-specific instance examples [Hois et al., 2006].

In summary, the vision system is specifically designed for handling uncertain information, maximizing information gain, and reasoning based on ontological queries over belief values. The ontological representation of the *PEO* module provides an ontological grounding of the sensomotoric features from the vision component and a general abstraction layer for the visual low-level processing. Instantiating the data from the vision component supports the analysis of scenes based on the visually classified objects. Although the classification results show high performance, the instance samples are too limited for a general performance evaluation and too constrained to the assignments by expert knowledge. In the next section, we present an approach for combining the *PEO* module with statistical visual data of a larger image corpus that omits expert knowledge and performs ontological queries over the TBox.

6.3.2 Indoor and Outdoor Scene Recognition

The second application scenario integrates an ontological representation for indoor and outdoor environments to determine the scene type [Reineking et al., 2009]. Similar to the use case above, single objects are visually classified and analyzed with regard to the overall scene type they imply. However, the visual system and the overall architecture as well as the domain and the ontological requests differ. As illustrated in figure 6.9, the scene classification system consists of four components: a domain ontology module for scene-specific information, a statistical model for scene-specific data, a reasoning module for actively updating the scene belief by minimizing uncertainty, and an image processing module consisting of type-specific object detectors.

Whenever an object detector is invoked, its positive or negative response induces a belief for the presence of the respective object type in the current scene depending on the estimated recognition rate of the detector. This object type belief is combined with the object-scene knowledge obtained from the domain ontology and the statistical model for updating the scene type belief. The domain ontology defines the vocabulary

---

[5]The *PEO* module ABox is available at `http://www.informatik.uni-bremen.de/~joana/ontology/modSpace/PhysicalEntityForActiveVision.owl`

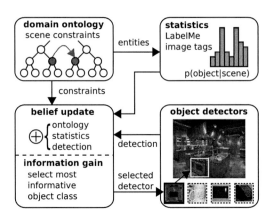

Figure 6.9: Scene Classification Architecture. The overview of the system architecture shows the four main components of the scene classification system: Object detectors (in the lower right part) detect and classify single objects within a scene. These results are analyzed with regard to scene constraints given by the domain-specific ontological representation and statistical data of the domain in order to yield the overall scene classification. (Figure taken from Reineking et al. [2011])

of object and scene types as well as relations between the two (e.g., kitchens contain cooking facilities). The statistical model provides co-occurrence probabilities of object and scene types which are learned from annotations available in the LabelMe image database [Russell et al., 2008]. In order to combine the uncertainty resulting from the object detection and from the statistical model with the set-based propositions from the ontology, we use Dempster-Shafer theory as it allows the assignment of belief values to arbitrary sets of propositions. Thus, it is capable of representing both uncertainty and logical constraints.

Based on a current scene belief, the system selects the most informative object type for the subsequent detection, i.e., it selects the object type which minimizes the uncertainty about the scene classification. This minimization is based on the support of object types to discriminate a scene. For instance, the object type 'stove' provides a high discrimination when the scene classification indicates kitchen or office. However, the minimization is also affected by the *recognition rate* of the respective object detector, e.g., if stoves are hard to recognize they provide only low discrimination. Each object detector is trained on a large set of sample images using boosting to obtain a strong binary classifier, which is systematically applied to different image regions during the detection phase. In addition, each detector is evaluated on a separate data set for estimating its recognition rate [Reineking et al., 2011].

In summary, the overall architecture consists of high-level classification and low-level visual analyses for the visual recognition. The bottom-up object detection is followed by a belief update based on statistical and logical inference. The top-down uncertainty minimization selects the next object type detection. In order to classify a scene, the system first computes the expected scene uncertainty reduction associated with searching for a specific object type in the scene. This is particularly useful if the context induces a prior belief that allows us to ignore less relevant objects from the beginning. After selecting the most informative object type, the vision module invokes the corresponding object detector and updates the current scene belief depending on the detection result, the constraints defined by the ontology for this object type, and the co-occurrence probabilities of the statistical model.

Similar to the use case in the previous section, we reuse the *PEO* module for specifying the domain-specific information, which represents general physical entities also available in the LabelMe database. Complementary to the statistical data analysis over the LabelMe database, we adjust the *PEO* module to refine the category restrictions for the representation of scenes and objects, i.e., the ontology module primarily formalizes the types of scenes and objects that exist in the domain as well as their relationships. Also, the ontological representation does not rely on a sample set of statistical data, but on expert knowledge and general common-sense knowledge of the domain. In contrast, statistical information about scenes results from an empirical process, namely the image annotations available in the LabelMe database. Thus, our aim is to combine both statistical data and ontological representation to improve the visual recognition: statistical information provides the probabilistic correspondences between objects and scenes obtained from a finite data set; the *PEO* module refinement provides not only logical constraints among scenes and objects of the domain but also the vocabulary for the statistical data.

As in the application scenario above, we reuse the *PEO* module structure for the domain modeling of scenes. As we define LabelMe-specific constraints, the refinement introduces new categories that are subsumed under dolce:non-physical-object to classify

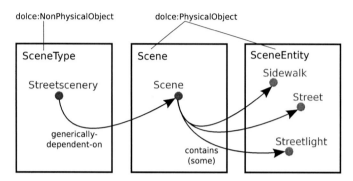

Figure 6.10: Scene Type Specification for Visual Recognition. Subcategories of SceneType are constrained by being related to a Scene that has to contain particular subcategories of SceneEntity. For example, an instance has to be related to an instance of Scene that contains at least one of the instances of Sidewalk, Street, or Streetlight, to be classified as Streetscenery.

scenes and dolce:physical-object to classify objects contained in scenes. Their dependencies are, however, bound to their appearances in samples from the LabelMe database as well as common-sense relations. Consequently, scenes not only present interior room scenes as in the previous use case but random types of scenes. Therefore, the *PEO* refinement introduces the categories SceneType and SceneEntity for scene and object types respectively. In essence, we use the same ontological structure for scenes and objects as in the previous application scenario; however, we can easily adjust the concrete dependencies between scenes and objects for the application scenario of classifying visual images taken from the LabelMe database.

The category SceneType specifies indoor as well as outdoor scenarios, which occur in the dataset. The *PEO* refinement particularly introduces the subcategories Bathroom, Bedroom, Kitchen, Livingroom, Mountainscenery, Office, and Streetscenery to represent the respective LabelMe sample data. Instances of these categories depend on concrete instances of the category Scene by the relation dolce:generically-dependent-on, which is similar to the representation of interior rooms in the previous section. These Scene instances can contain different objects classified as subcategories of SceneEntity. An example is illustrated in figure 6.10 and the respective OWL specification of the *PEO* refinement is presented in listing 6.4.[6]

Listing 6.4: Specification example from a *PEO* module refinement for vision recognition

Annotations :
> *rdfs : comment*   *' The PEO module refinement introduces new categories subsumed under DOLCE—Lite categories. The example illustrates the dependency between a subcategory of SceneType, its related Scene and the SceneEntities that are contained in the respective Scene. ' @en*

Class :   Streetscenery

  SubClassOf :
> SceneType ,
> generically —dependent—on **only** (Scene **and** (contain **some** Sidewalk)) ,
> generically —dependent—on **only** (Scene **and** (contain **some** Street)) ,
> generically —dependent—on **only** (Scene **and** (contain **some** Streetlight))

  DisjointWith :
> Bedroom ,
> Kitchen ,
> Office ,
> Livingroom ,
> Mountainscenery ,
> Bathroom

Hence, for every specific SceneType there is a number of subcategories of SceneEntity that necessarily have to occur in the Scene related to the SceneType. The distinction being drawn between SceneType and SceneEntity is based on the vision-centered perspective on the domain of possible scenes from the LabelMe database. Instances of SceneEntity (e.g., chair, refrigerator, or sink) are on the same level of granularity, whereas instances of SceneType (e.g., kitchen, street corner, or warehouse) are on a broader level of granularity and they particularly depend on a collection of the former. The levels of granularity depend on the vision system, which perceives its environment by analyzing individual entities in scenes (i.e., images from the LabelMe database), and the ontological representation reflects these granularity differences. As a result, the ontological representation follows the visual perception of the system and is thus inspired

---

[6]The *PEO* refinement is available at `http://www.informatik.uni-bremen.de/~joana/ontology/modSpace/SceneOntology.owl`

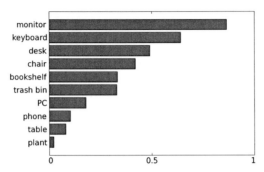

Figure 6.11: Example of Scene Type with Co-occurrence Disctribution. The distribution shows the average occurrence of scene entities that were identifiable to be contained in office scenes. (Figure taken from Reineking et al. [2011])

by vision-based embodiment [Vernon, 2008]. In particular, the *PEO* refinement takes into account precisely the objects that are classifiable by the system. Therefore, it currently specifies 7 different scene types and 24 different scene entities (as ontological representations follow the open world assumption, however, further scene types and scene entities are not excluded).

As outlined above, the ontological representation is complementary to the statistical model of the LabelMe database. In brief, the statistical model represents the relation of a scene class and an object class by their co-occurrence probability. A sample of 9601 scenes, which excludes those scenes from the LabelMe database that do not contain any known scene entities, together with 28701 known objects were used for the statistical analysis [Reineking et al., 2011]. An example of the co-occurrence distribution for the scene type Office is shown in figure 6.11. To compute the overall belief of the scene type for a given scene (image) from the dataset, knowledge about scene-entity relations from the statistical model and information from the *PEO* refinement are combined with the object detection responses from the vision process. In contrast to the use case in the previous section, no belief values are integrated into the ontology module but Bayesian probabilities are used for the statistical model and object detection. The statistical data for scene-entity relations is used for selecting visual detectors on scenes (cf. figure 6.9); the ontological constraints indicate the overall scene type on the basis of detected entities.

The scene type classification is based on the ontologically inferred scene type from detected objects and the estimated presence of objects from detector responses. For the evaluation, we use a set of detectors that always give correct responses based on the annotations in the LabelMe database. Together with the ontological specification of 7 scene types and 24 scene entities, we use all available scenes from the database annotated as one of these scene types if they contain at least one of the defined scene entities, resulting in 3824 scenes total. Given this setup, the performance of the scene recognition resulted in 94.7%, which indicates that the combined models accurately describe the domain. A visual recognition of a scene is classified correctly, if the actual scene type is assigned the highest plausibility value. However, the recognition performance varied among scene classes. For instance, 94.3% of offices and 84.0% of living rooms were classified correctly. This deviation is caused by some scene types having more discriminative scene entities than others.

Although the statistical model alone achieves a similar overall recognition performance, as the statistical analysis was performed on the same data set used during the training and recognition task, it is limited regarding the number of scenes for some classes. For instance, only 25 instances of the scene type bedroom are available in the LabelMe dataset sample. Thus, the statistical model is not expected to generalize well. The *PEO* module alone hardly ever achieves a similar performance, as it specifies only general (coarse) constraints between scene entities and scenes. For instance, the scene type office contains furniture and electronic equipment as general properties. However, the ontological representation induces a strong prior by restricting the set of possible scene types to a smaller subset of similar categories, e.g., different indoor scenes. This subset always contained the correct scene type.

In summary, both vision recognition applications benefited from the ontological representation by using their terminology to train low-level features or infer statistical data in order to result in an overall scene classification based on the ontological representation. The *PEO* module also provides sufficient structural information to ground the vision-specific categories into the ontological representation and to specify domain-specific refinements. Both application scenarios also showed the useful reuse and refinement mechanism by using the *PEO* module to satisfy domain-specific requirements of scenes and objects they can contain. Both systems also resulted in high performance for the overall vision recognition process. The first use case allowed a fine-grained specification of ABox instantiations and their functional characteristics together with belief values, however, the domain was limited. The second use case allowed a general specification of an, in principle, unlimited domain, however, statistical distributions over scene-entity relations were required for distinct results. In conclusion, the integration of spatial domain-specific ontology modules in visual systems provides an abstract knowledge layer to the system that can further be used together with other applications, such as robotic systems or scene-to-text generation systems.

6.4 Spatial Language Interpretation

The interpretation of spatial language is essential for tasks in areas related to Artificial Intelligence, such as robotics [Roy and Reiter, 2005], human-computer interaction [Moratz and Tenbrink, 2006], computational linguistics [Kruijff et al., 2007], the Semantic Web [Fonseca and Rodriguez, 2007], or Geographic Information Systems [Mani et al., 2008]. In these areas, different formalisms have been developed that address the problem of interpreting natural language expressions in terms of their situation-based or contextualized meaning. The formalisms being used differ across fields. In GIS, for instance, methods for spatial language understanding are often based on formal symbolic models [Grütter et al., 2010; Mani et al., 2008], whereas in robotics and computational linguistics methods prevail that are based on corpus-based, statistical, or sensor data [Kelleher and Costello, 2009; Roy and Reiter, 2005]. Many approaches are also inspired by empirical findings and psycholinguistic studies in spatial cognition research (e.g., Talmy [2006]; Carlson and van der Zee [2005]; Zlatev [2007]) with the aim to improve spatial language understanding by imitating the way humans seem to understand and use spatial language.

The situational interpretation of spatial language relies on various factors and results in a rather complex task, in particular, as spatial information in natural language can imply multiple meanings and induce different spatial constraints on a spatial situation. Even the meaning of a 'simple' linguistic "in front of" relation needs an in-depth analysis of its potential interpretations and influencing aspects:

> So we see that the interpretation of 'in front of' seems to depend on
> complex social conventions about object type, as well as on many issues of
> physical simulation. (...) Similar considerations lead[sic] us to realize that

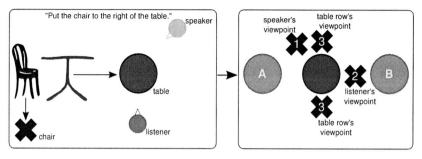

Figure 6.12: Linguistic Utterance Example. The utterance "Put the chair to the right of the table." can be spatially interpreted in different ways depending on locations of speaker and addressee, frame of reference, context, and dialogue history. The right side shows four possible interpretations: the chair position 1 is interpreted as the 'right-of-the-table' from the intrinsic viewpoint of the speaker, chair position 2 is interpreted as the 'right-of-the-table' from the intrinsic viewpoint of the listener, both chair positions 3 are interpreted as the 'right-of-the-table' in a row of other tables, such as A and B, i.e., positions 3 are interpreted as seen from the relative viewpoint of tables rows in either of both directions (from A to B or B to A).

collision detection and path planning are relevant to spatial expressions involving action verbs, and that many issues of friction and complex object geometry, for example, are relevant to prepositions dealing with placement. [Smith, 2000, p. 129]

Figure 6.12 illustrates an example, in which the utterance "Put the chair to the right of the table" can have multiple meanings depending on the spatial frame of reference (cf. section 4.2.4.1), which is not expressed in the utterance itself. To derive the spatial situation by means of the linguistic expression, not only the individual linguistic features that contribute to the spatial meaning have to be determined, but also external, non-linguistic criteria and possible alternative meanings have to be taken into account. Kracht [2008], for instance, identifies eight parameters necessary to interpret natural language, namely the spatial language expression (utterance), the speaker, the addressee, the utterance time, the story time, the speaker's location, the addressee's location, and the vantage point [Kracht, 2008, p. 100].

Hence, for an adequate approach to spatial language interpretation not only information given by the linguistic expressions itself but also situation-specific (non-linguistic) elements need to be disentangled and modeled individually in order to clearly specify their dependencies and interactions. In essence, "the cleaner the separation achieved between the linguistic semantics of space and the non-linguistic, situation-specific interpretation of space, the more flexible and adequate an approach to spatial language can become" [Bateman, 2010b, p. 46]. From an ontological viewpoint, the different types of information that contribute to interpret a spatial language expression can be distinguished according to their spatial perspectives presented in section 3.1: The linguistic utterance can be modeled by ontological modules complying with a multi-modal spatial perspective for language-related spatial information; situation-dependent information about the speaker, the addressee, the utterance time, and the story time can be modeled by ontological modules with abstract and domain-specific perspectives; and information about locations and vantage points can be modeled by modules with qualitative and quantitative perspectives.

Our distinction of the spatial perspectives, and accordingly the spatial ontology modules, thus implies the following module separations:

Linguistic Perspective. The modeling of spatial utterances in a linguistic module needs to reflect spatial information available in the language. The module consequently has to specify the main spatial linguistic aspects, such as trajectors (the entities whose locations or positions are described), landmarks (the reference entities in relation to which the locations or the motions of the trajectors are specified), relations (constraints on spatial properties of trajectors or landmarks), motions (movements of spatial entities), or paths (start, intermediate, and end points of motions) [Zlatev, 2007; Talmy, 1983]. A resulting ontology module that models these aspects and their constraints has been presented in section 4.2.4.1, namely the *GUM-Space* module.

Abstract and Domain-Specific Perspective. The modeling of situations and contextual information in a domain-specific module can reuse abstract modules to provide them with high-level descriptions of specifying situational aspects and the domain-specific modules can be specialized for application-specific requirements. For instance, the representation of sensor data or environmental maps and object types in the environment that is used by a human-computer interaction system can be modeled in domain-specific ontological modules. Having a representation for linguistic information and domain-specific information, both representations can be related with each other to interpret the linguistic expressions in terms of their domain entities [Barclay and Galton, 2008]. This can technically be achieved by combining the ontological modules and by defining their module connections.

Quantitative and Qualitative Perspective. The modeling of spatial arrangements needs to represent the spatial locations of entities and their spatial relations, which can be achieved by using a quantitative module that may reflect application-specific data formats or by using a qualitative module that may reflect abstract spatial representations [Tellex, 2010]. For instance, the *RBO* module that has been presented in section 4.2.1.1 can be used for region-specific relations and requirements for spatial locations. The underlying qualitative region-based meaning of a linguistic utterance can then be reflected using such an ontological module.

Two application scenarios that we present and evaluate in this section deal with modeling the linguistic and the qualitative spatial perspectives and their combinations in order to provide a method for spatial language interpretation. We address the problem of mapping raw linguistic data to qualitative spatial representations, i.e., how natural language expressions can be interpreted in terms of a qualitative spatial 'layout' of the situation. The next section introduces this relationship on a theoretical level for modeling the module combinations, the subsequent section shows empirical results of combining language with qualitative spatial representations.

6.4.1   Connection of Modules for Spatial Language Interpretation

As we understand the (situational) interpretation of spatial language as a translation from the linguistic *GUM-Space* module to other spatial modules, we investigate the specification of module combinations that include the *GUM-Space* module. In particular, we interpret spatial language in terms of spatial qualitative representations. These types of module combinations are intended to map spatial language to a spatial qualitative formalism by reusing relevant spatial modules presented above and combinations between them. For this purpose, we developed perspectival $\mathcal{E}$-connections, an approach that has been introduced in Hois and Kutz [2008b], and that has been extended in Hois and Kutz [2008a]. This approach defines $\mathcal{E}$-connections between spatial modules that comply with the spatial perspectives developed in this thesis.

Here, we focus on link relations that are defined between the *GUM-Space* module and modules complying with qualitative spatial perspectives in order to provide an interpretation of spatial language. This particular modeling of spatial language interpretation is motivated by making the distinction between language and space transparent, as it is this distinction that "allows for the specification of a semantic representation that captures what the linguistic content of spatial expressions is committing to" [Bateman, 2010a], that separates the linguistic content from the formal spatial representation, and that provides a flexible way to combine language with other spatial formalisms [Hois and Kutz, 2008b].

The overall aim is to provide a general framework for identifying links between language and space as a generic approach to spatial communication, which is not limited to concrete kinds of applications in which it is used. In particular, rather than attempting to integrate the most general spatial theories, we use various specialized spatial modules that comply with the qualitative perspective to be able to support dedicated and optimized reasoning. Technically, this is represented as a structured logical theory in CASL [Mossakowski et al., 2008]. First, however, we start with some examples to outline the idea and subsequently we show the formalizations of the $\mathcal{E}$-connections between the modules.

Figure 6.13 illustrates four example interpretations of qualitative spatial representations in the '9^+-intersections for topological relations between a directed line segment and a region'-calculus (abbreviated here as 9^+-intersection calculus). The 9^+-intersection calculus is a qualitative spatial representation that relates a direction with a region [Kurata and Egenhofer, 2007]. It distinguishes between the starting and end point of the line as well as the interiors, boundaries, and exteriors of the region. The 9^+-intersection calculus specifies 26 types of basic relations that a directed line and a topological region can have under these constraints. Directed lines represent movement patterns with regard to regions. The direction of the lines indicates the start and end point of the movement of a moving entity with respect to a specific region. A movement may, for instance, start from the exterior of the region and end at its boundary.

The example sentences in figure 6.13, which describe motion configurations with regard to a reference object, can be represented with the elements in the 9^+-intersection calculus. The path of the entity that moves is represented as the directed line; the reference object in relation to which the motion is described is represented as the topological

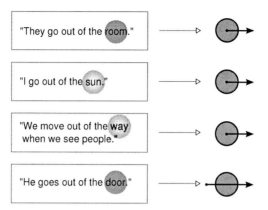

Figure 6.13: Language to Space Mapping Examples. Four example sentences, which have identical linguistic structures in the *GUM-Space* module, are represented by similar structures in the qualitative module. However, exact mappings of linguistic entities to regions and motions rely on additional perspectives.

region. In the first sentence, the linguistic instance 'room' is mapped directly to the spatial region and the motion of the linguistic instance 'they' is mapped to the directed line that has its start inside and end outside the region. In the second sentence, the linguistic instance 'sun' is mapped to the spatial region via an object-specific function that determines the respective sunny area and the motion of the linguistic instance 'I' is mapped to the directed line that has its start inside and end outside the region. In the third sentence, the linguistic instance 'way' is mapped to the spatial region via a context-specific function that determines the corresponding area and the motion of the linguistic instance 'we' is mapped to the directed line that has its start inside and end outside the region. In the fourth sentence, similarly to the first sentence, the linguistic instance 'door' is mapped directly to the spatial region and the motion of the linguistic instance 'he' is mapped to the directed line that, unlike the other example sentences, starts and ends outside the region but is inside the region during the motion.

Hence, a mapping from the linguistic entities to the qualitative entities cannot be achieved directly. Although the example "They go out of the room." can map the motion of 'they go' to the directed line and the reference object 'room' to the region, the examples "I go out of the sun." and "We move out of the way when we see people." need more specific mapping functions to connect the 'sun' and the 'way' with the topological region, which requires the integration of domain-specific modules. However, the intended meaning of 'going out of' may not always be interpreted as the same basic relation in the 9^+-intersection calculus. The example "He goes out of the door." shows that the directed line is related to the topological region in a different way than in "They go out of the room.", even though the instances 'door' and 'room' can be related directly to the region. Again, a domain-specific module that provides property information for door and room entities can support determining the concrete connection between the linguistic and qualitative modules.

As the example sentences imply that a separation between the different spatial perspectives is necessary to analyze their varying influences on the actual interpretation, a formal relationship between the linguistic module and the qualitative spatial module has to be defined in order to provide for their flexible interplay. As we combine two modules with distinct spatial perspectives, namely a linguistic and a qualitative directional and topological spatial module, link relations have to be used as a combination method between the two thematically distinct ontological modules. In the examples, the linguistic module *GUM-Space* needs to be connected with the qualitative module of the 9^+-intersection calculus, which results in a heterogeneous interface ontology that contains the relations between spatial language and qualitative space. This perspective-driven connection is illustrated in figure 6.14, in which parts from both modules are combined by using $\mathcal{E}$-connections between the individual parts. For instance, the linguistic instance 'B' is linked to a topological region, the linguistic instance 'A' is linked to a directed line, and the linguistic spatial modality 'DenialOfFunctionalControl' is linked to a specific 9^+-intersection base relation.

In this scenario, the resulting interface ontology only formulates the mapping from language to space, i.e., if a particular ontological construction is instantiated in the linguistic module its link relation to the qualitative module can be formulated. Although the reverse case can also be specified in the same or another interface ontology, which produces a mapping from space to language that can be used for the generation of natural language, it would depend on other contextual and object-specific aspects that may lead to diverse linguistic expressions [Kelleher and Kruijff, 2006]. Consequently, the generation of spatial language by means of a mapping from space to language is

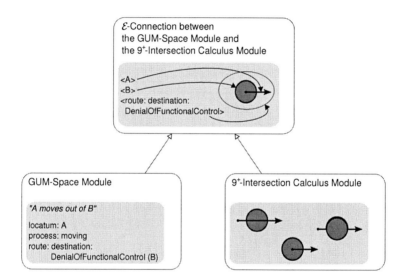

Figure 6.14: Interface Ontology for Spatial Language and Qualitative Space.
The connection between a multimodal (linguistic) and a qualitative (directional and topological) spatial module results in a heterogeneous interface ontology module. Gray boxes illustrate instance examples.

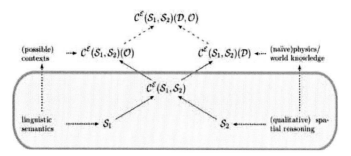

Figure 6.15: Perspectival $\mathcal{E}$-Connections for Spatial Language Interpretation.
Structured modules provide perspective-dependent information that contributes to the
overall meaning and interpretation of spatial language. The box highlights the spatial
modules and their combination on which we focus in this section.

not the set of inverse link relations of those used to define the mapping from language
to space.

The entire approach to connect different modules to allow spatial language interpretation is illustrated in figure 6.15. Both modules for linguistic and qualitative
spatial information are connected by an $\mathcal{E}$-connection that yields the interface ontol-
ogy $\mathcal{C}^{\mathcal{E}}(\mathcal{S}_1, \mathcal{S}_2)$. This module can be further extended by adding contextual or world-
knowledge from domain-specific ontology modules. A combination of these extensions
allows the interpretation of spatial language and it represents the theory of *Perspectival $\mathcal{E}$-Connections* [Hois and Kutz, 2008b], in which multiple modules with different
perspectives are combined in order to provide an overall meaning and interpretation.

The modules for linguistic and qualitative spatial information have been specified by using the specification and ontology language CASL. Whereas the examples
above combine spatial language with the 9^+-intersection calculus, we now present the
connected ontological specification by combining GUM with the double-cross calculus
(DCC) [Freksa, 1992], a ternary calculus for spatial orientations. DCC distinguishes
15 relations between an observer at a position A, who is oriented (or moves) towards
an entity at a position B (cf. figure 6.16). The 15 orientation relations are defined
along three axes motivated by human cognitive characteristics: A and B are called the
perspective point and the *reference point* respectively in Renz and Nebel [2007]. They
determine the front-back axis. Orthogonal to this axis are two further axes specified by

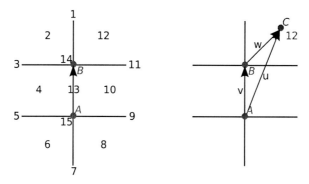

Figure 6.16: Double-Cross Calculus. The left figure shows the 15 qualitative orientation relations of the ternary DCC, which are based on two points (A and B) and their axes according to Freksa [1992]. The right figure shows an example, in which a third point C has the qualitative spatial relation 12 with respect to the axis based on the points A and B.

A and B. Another entity located at some position C can then be described according to one of the 15 orientations. The arrows that result from the connections among A, B, and C are referenced as v, w, and u (cf. figure 6.16).

As the DCC and its composition table is specified in CASL[7], it is linked with the CASL version of the *GUM-Space* module to interpret static spatial descriptions with projective relations between located objects and reference objects. In detail, the linguistic instances that have to be mapped to the qualitative representations are the located object, the spatial relation, the reference object, and the frame of reference. They are specified in *GUM-Space* by the relations and categories locatum.SimpleThing, hasSpatialModality.SpatialModality, relatum.SimpleThing, and spatialPerspective.Element (even though the spatial perspective is often not linguistically expressed). The sentence "The car is in front of the house." (assuming a relative frame of reference) is thus instantiated as shown in listing 6.5.

Listing 6.5: *GUM-Space* instantiation of "The car is in front of the house."

Individual: StaticDescriptionSentence
    Annotations:

---

[7]https://svn-agbkb.informatik.uni-bremen.de/viewcvs/Hets-lib/trunk/Calculi/Space/
OrientationCalculi.het (visited on July 21, 2009)

>   rdfs:comment   'Instantiation of the whole sentence:
>   *The car is in front of the house*.' @en
> **Types**:
>   SpatialLocating
> **Facts**:
>   gum: processInConfiguration   BeingProcess,
>   locatum   TheCar,
>   placement   GeneralizedLocationFrontOfHouse

Individual: BeingProcess
>   **Annotations**:
>   rdfs:comment   'Instantiation of the process:
>   *is*.' @en
>   **Types**:
>   gum: Process

Individual: TheCar
>   **Annotations**:
>   rdfs:comment   'Instantiation of the locatum:
>   *The car*.' @en
>   **Types**:
>   gum: SimpleThing

Individual: GeneralizedLocationFrontOfHouse
>   **Annotations**:
>   rdfs:comment   'Instantiation of the placement:
>   *in front of the house*.' @en
>   **Types**:
>   GeneralizedLocation
>   **Facts**:
>   hasSpatialModality   FrontProjectionOfHouse,
>   relatum   TheHouse

Individual: FrontProjectionOfHouse
>   rdfs:comment   'Instantiation of the spatial modality:
>   *in front of*.' @en
>   **Types**:
>   FrontProjectionExternal

Individual: TheHouse
>   rdfs:comment   'Instantiation of the relatum:
>   *the house*.' @en
>   **Types**:
>   gum: SimpleThing

The individual instances that are used in such static directional descriptions are mapped to the available entities in the DCC, i.e., the points and their relations. Clearly, the instances for the locatum and relatum are mapped to the points and the spatial modality is mapped to the qualitative relation of the DCC. More precisely, we specify the interface ontology module *GUMwithDCC*[8] as a result of the $\mathcal{E}$-connections between the linguistic *GUM-Space* module and the qualitative *DCC* module. The *GUMwith-DCC* module connects both input modules and defines their link relations, namely the link relations referenceOfModality, modalityToOrientation, and simpleThingToLocation are introduced in *GUMwithDCC*:

referenceOfModality. This link relation maps a SpatialModality to a FrameOfReference. It thus relates the spatial modality from the *GUM-Space* module with the reference frame from the *GUMwithDCC* module. The category FrameOfReference can be one of the disjoint types intrinsic, relative, or absolute. It is used in the *GUMwithDCC* module to distinguish between external and internal directional relations, i.e., whether the directional relation is described from an intrinsic, relative, or absolute perspective, which leads to different mappings. Often, the frame of reference is not directly available from the linguistic expression, and although it is relevant for the actual mapping, determining the reference frame may have to be solved by contextual or external information. The next section also provides a solution to use statistical data that allows us to ignore the reference frame.

modalityToOrientation. This link relation maps a SpatialModality to an OrientationDCC. It relates the spatial modality from the *GUM-Space* module with one of the 15 orientation relations in the *DCC* module. The relation allows an instance of SpatialModality to be mapped to multiple double-cross orientations because a linguistic expression may use a directional term that can be interpreted by more than one orientation. For instance, in the sentence "The car is parked to the left of the tree (from my perspective).", the instance of LeftProjectionExternal in the *GUM-Space* module that specifies the directional relation can be mapped to the orientations 2, 3, and/or 4 in the *DCC* module, mapping 'car' to C, 'tree' to B, and '(my)' to A (cf. figure 6.16 for the points and relations).

---

[8]Available at `https://svn-agbkb.informatik.uni-bremen.de/viewcvs/Hets-lib/trunk/Ontology/GUM/test/GUMFullwithDCC.casl` (visited on July 1, 2010)

simpleThingToLocation. This link relation maps a SimpleThing to a Location. It relates the linguistic instances for the located object and the reference object in the *GUM-Space* module to points in the *DCC* module. The relation is functional, i.e., every linguistic entity can only be mapped to one point in the *DCC* module.

Listing 6.6 shows a part of the *GUMwithDCC* module specification, in which a BackProjection from the *GUM-Space* module is mapped to the respective orientations. In the *DCC* module specification in CASL, the 15 relations have individual names instead of the 15 numbers. back refers to orientation 7, leftBack refers to orientation 6, rightBack refers to orientation 8, front refers to orientation 1, leftFront refers to orientation 2, and rightFront refers to orientation 12. The listing examples show, that the frame of reference determines potential mappings. If the reference frame is intrinsic, the perspective is derived from the position of point A in the double-cross calculus. Thus, an expression such as "The cat is behind me." results in a mapping from the linguistic spatial modality of BackProjection to the orientations 6, 7, or 8.

Listing 6.6: Mapping of a BackProjection in the *GUMwithDCC* module

```
forall bp: BackProjection
    %% the frame of reference determines the mapping %
    %% to the DCC relations for instances of BackProjection %
    . referenceOfModality(bp) = intrinsic   =>
            modalityToOrientation(bp) = back
        \/  modalityToOrientation(bp) = leftBack
        \/  modalityToOrientation(bp) = rightBack

    . referenceOfModality(bp) = relative    =>
            modalityToOrientation(bp) = front
        \/  modalityToOrientation(bp) = leftFront
        \/  modalityToOrientation(bp) = rightFront
```

The specification of *GUMwithDCC* is formulated in CASL because the *DCC* module and also a version of the *GUM-Space* module are formulated in this language. As heterogeneous specifications in CASL can also contain Manchester OWL syntax, the OWL version of *GUM-Space* module could have also been used here.

The example mapping in listing 6.7 reflects the specification of the linguistic expression "The tree is to the left of the house (from my point of view)." and maps the linguistic instances to the respective locations and orientations. The proof in the specification shows that respective orientations as specified in the *GUMwithDCC* module can be inferred for a concrete linguistic expression. Here in particular, the orientation 2 (leftFront) in the double-cross specification is a valid orientation for the linguistic expression.

Listing 6.7: Mapping example from language to space in the *GUMwithDCC* module

```
ops
    %% instantiation of linguistic entities: %
    %% ''The tree is to the left of the house %
    %% (from my point of view).'' %
    sl                 :  SpatialLocating;
    house, tree, me :  SimpleThing;
    gl                 :  GeneralizedLocation;
    lp                 :  LeftProjection;
    u, v, w            :  Arrow;

    %% specification of relations among linguistic entities %
    . not(me = house)
    . not(me = tree)
    . not(house = tree)
    . locatum(sl, tree)
    . placement(sl, gl)
    . relatum(gl, house)
    . hasSpatialModality(gl, lp)

    %% instantiation of the arrows between the three points in DCC %
    . v = simpleThingToLocation(me)   —>  simpleThingToLocation(house)
    . w = simpleThingToLocation(house) —>  simpleThingToLocation(tree)
    . u = simpleThingToLocation(me)   —>  simpleThingToLocation(tree)

    %% proof that the orientation leftFront (orientation 2 in DCC) %
    %% is one possible orientation for the linguistic instantiations %
    %% (using the CASL implies statement) %
then %implies
    %% -- the arrow v needs to go from *me* to *house* %
    %% -- the arrow w needs to go from *house* to *tree* %
    %% -- the e-connection modalityToOrientation needs to link %
    %% lp with leftFront %
    %% -- the resulting arrow u (v#w) needs to be leftFront %
    . v = simpleThingToLocation(me)   —>  simpleThingToLocation(house)
    /\ w = simpleThingToLocation(house) —>  simpleThingToLocation(tree)
```

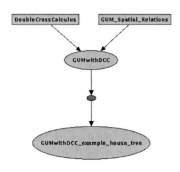

Figure 6.17: *GUMwithDCC* Proof. The left diagram in HETS shows that the *GUMwithDCC* module imports the linguistic and the qualitative modules and an example extends this specification. The orientation-based interpretation of "The tree is to the left of the house" is instantiated in the GUMwithDCC_example_house_tree specification. The red dot indicates unsolved proof obligations in this theory. These are the proof implications shown in listing 6.7. The right diagram shows that the implications were proved to be correct using the SPASS theorem prover.

```
/\ locatum(sl, tree) /\ placement(sl, gl)
/\ relatum(gl, house) /\ hasSpatialModality(gl, lp)
/\ modalityToOrientation(lp) = leftFront %(linguistic leftfront)%
. v # w > leftFront %(proof 1 leftfront)%
. (v|w=leftFront /\ v|u=leftFront) %(proof 2 leftfront)%
```

The proof implication by the specification in listing 6.7 can be shown to be correct in the system HETS [Mossakowski et al., 2007]. Figure 6.17 shows the result when analyzing the implications with the prover SPASS [Weidenbach et al., 2009] in HETS. The diagrams show the proof of the correctness of the linguistic example and its orientation implications. This also shows that the $\mathcal{E}$-connections between the *GUM-Space* module and the *DCC* module as specified in the *GUMwithDCC* module are able to infer the qualitative orientations given the linguistic instantiations. It also shows that spatial language interpretation in terms of a mapping from spatial language to qualitative spatial representations can be supported by using this specification.

As shown above, more contextual aspects are necessary for a final mapping from linguistic spatial relations to qualitative models. The next section presents an empirical mapping from language to space by using corpus data and statistical methods in order to avoid the difficulty of specifying further contextual aspects (which may not always be available or inferable).

6.4.2 Empirical Data for Connecting Language and Space

The previous section presented a method for spatial language interpretation by manually specifying link relations between linguistic and qualitative modules. However, link relations can also be automatically derived from corpus data. In particular, learning techniques can be applied for measuring the likeliness of qualitative spatial relations for certain linguistic constructions. This method has the advantage that no additional context or world knowledge needs to be specified and to be available for the spatial language interpretation process. However, it also causes uncertainty in the mappings as qualitative relations for individual linguistic expressions can only be inferred with a certain probability or likelihood. Thus, we briefly present an analysis of the automatic learning process followed by a formal approach to model the resulting uncertainties.

To use machine learning algorithms for detecting correlations between spatial linguistic expressions and their qualitative spatial representations, corpus data that provides both linguistic and qualitative information is required for the learning phase. So far, only few approaches have attempted to collect corpus data that is sufficient to train the mapping from language to space. Kordjamshidi et al. [2013] show one such corpus with more than 2000 annotated sentences, containing annotations for linguistic features and qualitative relations. Linguistic corpora typically annotate only the linguistic (syntactic or semantic) features. For example, such corpora have been used to evaluate the adequacy of the *GUM-Space* module [Hois, 2010b].

The spatial (English) language corpus in Kordjamshidi et al. [2013] contains selections from the ImageCLEF IAPR corpus [Grubinger et al., 2006] for photographic scene descriptions, from the HCRC Map Task corpus [Anderson et al., 1991] for spatial routes, and from a room description corpus [Shi and Tenbrink, 2009]. Although many of the corpus entries (1401 sentences) contain static scene descriptions, there are also entries that do not contain space-specific language, and thus the corpus is biased but not entirely domain-specific, and non-spatial descriptions exist for reasons of generality. However, the predominating static expressions induce an annotation of qualitative spatial representation that can benefit from this distribution. Consequently, we use a static qualitative calculus, namely the RCC-8 (cf. section 2.6.2 for an introduction), to which we map the spatial language expressions.

In the corpus, every sentence has annotations for its located or moving object (trajector), spatial relation (spatial-indicator for the linguistic expression and spatial-modality for the *GUM-Space* modality type of the spatial relation), reference object (landmark), and the general and specific qualitative spatial types. The sentences also have annotations on whether they contain a static or dynamic description, path information, and a frame of reference. These latter annotations are, however, not relevant for the purpose of mapping language to RCC-8, as indicated by the results below. An annotation example for a sentence from the linguistic corpus is shown in table 6.1. The qualitative types that are distinguished refer to the general spatial type of relation (general-type), e.g., region, distance, orientation, or shape, the specific type that can be used (specific-type), such as a specific qualitative calculus, and its concrete value in that calculus (spatial-value). As the example shows, annotations can have more than one qualitative type. In detail, the corpus contains 1040 annotated *topological*, 662 *directional*, and 91 *distance* relations.

Table 6.1: Corpus Annotations for Linguistic and Qualitative Information. Every sentence entry in the linguistic corpus in Kordjamshidi et al. [2011a] has annotations for aspects of spatial language and qualitative relations.

"Two people are sitting next to her."	
trajector:	people
landmark:	her
spatial-indicator:	next to
spatial-modality:	Proximal
general-type:	region / distance
specific-type:	RCC-8 / relative-distance
spatial-value:	dc / close
dynamic:	static
path:	none
frame-of-reference:	none

Although we focus on learning the automatic mapping from linguistic annotations to the qualitative spatial types, it has also been shown that the extraction of the linguistic annotations can be done automatically with high enough reliability (precision and recall) by using machine learning techniques [Kordjamshidi et al., 2011b]. In the following experiments, however, we use machine learning to automatically infer the qualitative spatial relation from a linguistic expression as represented by its annotation. In particular, the learning uses sets of sentences from the annotated corpus to classify their respective RCC-8 relations. The prior distribution of the RCC-8 relations in the corpus is shown in table 6.2.

Table 6.2: Distribution of RCC-8 Relations. Overall RCC-8 annotations in the linguistic corpus in Kordjamshidi et al. [2011a].

RCC-8	dc	tpp	ntpp	ec	eq	ntpp^{-1}	tpp^{-1}	po	none	Total
#Corpus Entries	261	374	15	461	10	1	25	15	747	1909

As we have shown in Kordjamshidi et al. [2011a] and Kordjamshidi et al. [2013], best results for the automatic mapping from linguistic annotations to qualitative relations are obtained by using *support vector machines* (SVM) [Cristianini and Shawe-Taylor, 2002]. These provide a supervised learning technique, which uses a pre-defined training data set to learn data patterns and to classify them. Accuracy of their learning results is tested on a test data set. Here, SVM classifiers were trained with subsets of the annotated corpus sentences to classify RCC-8 relations. Accuracy is calculated by averaging over a 10-fold cross-validation of the corpus data, i.e., the data set is divided into 10 parts and trained 10 times, every time with a different test set and 9 training sets. The F-measure measures the classification performance by the harmonic mean of precision and recall and it is computed in the standard way [van Rijsbergen, 1979], as follows:

$$\text{recall} = \frac{\text{TP}}{\text{TP+FN}} \quad ; \quad \text{precision} = \frac{\text{TP}}{\text{TP+FP}} \quad ; \quad \text{F-measure} = \frac{2*\text{precision} * \text{recall}}{\text{precision} + \text{recall}}$$

with

TP = # correctly classified samples that belong to the annotated RCC-8 relation
FN = # samples incorrectly not assigned to the annotated RCC-8 relation
FP = # samples incorrectly assigned to the annotated RCC-8 relation

For mapping the linguistic annotations to RCC-8 relations, we conducted two main experiments and compared their average classification accuracies, i.e., their learning performance. Both experiments gradually add the linguistic annotations of the corpus as input features for the learning. The sequence of linguistic input features and their results are illustrated in figure 6.18[9]. The first experiment (figure 6.18, upper chart) starts with the spatial indicator (spi) to learn the classification, which yields a rather low F-measure of 0.474. Adding the landmark (lm) to the learning increases the F-measure to 0.753. Gradually adding the trajector (tr) annotation results in an acceptable and stable F-measure of 0.833. Adding the additional annotations for motion (mo), path (pa), static/dynamic (dy), and frame of reference (for) has only little effect on the final result of 0.877. Adding these annotations sometimes even decreases the F-measure value, which is caused by the combination of noise and lack of data.

The second experiment (figure 6.18, lower chart) starts the learning by using the spatial modality (mod) of the *GUM-Space* module. Using only this annotation as input features yields an F-measure of 0.719. This rather high accuracy may be a result of the high expressivity of *GUM-Space* and its in-depth exploration with regard to the spatial relations expressed in spatial language. Adding the spatial indicator to the learning has no significant effect on the performance, which increases accuracy by less than 0.02. Adding the annotations for the landmarks as input features increase the F-measure to 0.807, and gradually adding the input feature for the trajector yields an acceptable and stable F-measure of 0.878. The significant outcome of these experiments is that using the spatial modality as an input feature for the learning eliminates trivial fluctuations in the accuracy of the classifiers. The learning results are more robust against extra and noisy features.

However, the results are also affected by the prior distribution of the RCC-8 relations. As table 6.2 shows, there are only very few examples for some RCC-8 relations, in particular, ntpp^{-1} was only annotated once. This may be caused by a lack of variety in the corpus selection but also by the inadequacy of some RCC-8 relations to represent linguistic expressions. The classification of the eq relation, which has been annotated 10 times in the corpus, has an F-measure of 0.8 using all linguistic input features. When using fewer features, the F-measure decreases by 0.4. Using fewer input features, however, has a reverse effect on classifying the po relation, which has been

---

[9]The list of input features that are marked with an asterisk contain additional features that annotate full phrases of trajector, landmark, and spatial indicator. As they have little effect on the overall results, they have been omitted for reasons of clarity.

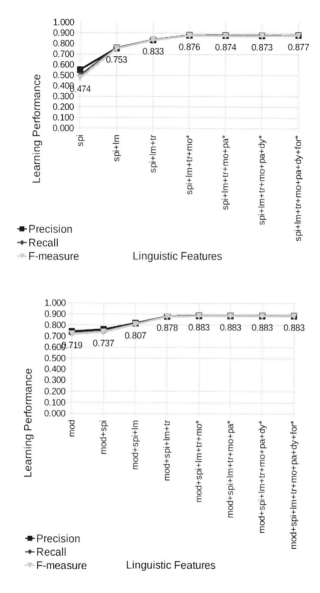

Figure 6.18: Machine Learning Results for Linguistic-Spatial Mapping. Learning performance (accuracy) of the RCC-8 SVM classifier, gradually adding linguistic input features. (adapted from Kordjamshidi et al. [2011a])

annotated 15 times in the corpus. Its F-measure increases by 0.11. This effect is caused by lack of sufficient data for annotated eq and po relations.

The stabilized F-measure for all classifications when learning with either spi+lm+tr in the first or mod+spi+lm+tr in the second experiment indicates that the other additional annotations have no significant impact on the accuracy. This is plausible as the corpus contains primarily static descriptions without motion or path information but also as motion-related entries may not be annotated with a specific topological RCC-8 relation. Consequently, these additional features are not discriminative for the learned RCC-8 relations. As the frame of reference is often annotated as 'none' (for topological relations) it did not have an effect on the performance. This can also indicate, as pointed out in the previous section, that the frame of reference (or other additional annotations) is not necessary for interpreting spatial relations between regions (or other qualitative relations).

Although these experiments were only carried out on static descriptions to learn 8 different topological relations, the results indicate that similar accuracies may be achieved for classifying directional, distance-, or shape-related qualitative types, if appropriate corpus data is available. In addition, annotations for contextual data or world knowledge may be used as an input feature for the learning. In this way, object-related properties can support the learning to infer the general qualitative type, e.g., whether the object can have an intrinsic reference frame and certain directional qualitative relations can be implied. For the spatial language interpretation task, the SVM classifiers can be used to derive concrete RCC-8 relations and to supply a general mapping relation from language to space that takes into account potential uncertainties. In particular, the perspectival $\mathcal{E}$-connections approach that has been presented in the previous section can be enriched by these uncertainty values.

In Hois and Kutz [2008a] and Hois et al. [2008], we have demonstrated that perspectival $\mathcal{E}$-connections can be extended with similarity values in order to take into account uncertainties when interpreting spatial language in terms of qualitative calculi. The similarity values reflect the likelihood that a given ontological instantiation in *GUM-Space* that reflects a natural language input can be mapped to a specific qualitative spatial relation. Hence, the value indicates how similar the respective parts from the two modules are to each other. Technically, the link relations that are specified or determined between instances from the *GUM-Space* module and a qualitative spatial

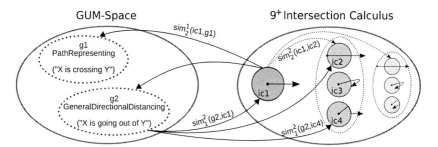

Figure 6.19: Similarities for Mapping Language to Space. Example of S-connections between the *GUM-Space* module and the 9^+-intersection calculus. Different similarity connections are outlined, such as $\mathsf{sim}_1^2(g2, ic1)$ for a connection from *GUM-Space* to the 9^+-intersection calculus, $\mathsf{sim}_2^1(ic1, g1)$ for a connection from the 9^+-intersection calculus to *GUM-Space*, and $\mathsf{sim}_2^2(ic1, ic2)$ for similarities between the 9^+-intersection calculus relations. (adapted from Hois et al. [2008])

module are extended by an actual similarity value. This type of uncertainty is not influenced by an agent's instantiation of ontologies but by the statistical distribution of related linguistic features and spatial relations. The resulting S-connections [Hois and Kutz, 2008a] approach can be used to specify the most likely related, i.e., similar, ontological constructs from *GUM-Space* and a qualitative calculus. An example of similarities that relates the *GUM-Space* modules with the 9^+-intersection calculus is shown in figure 6.19.

The example illustrates how such similarities can be described for connections across modules. In the *GUM-Space* module, spatial modalities can be connected with spatial relations in the 9^+-intersection calculus by defining a similarity value between the entities together with their connection. The two example categories PathRepresenting and GeneralDirectionalDistancing are related by such a connection with the 9^+-intersection calculus relation examples. In principle, the connection can be defined in both ways, i.e., the connection $\mathsf{sim}_1^2(g2, ic1)$ relates a spatial modality instance $g2$ in the *GUM-Space* module to the spatial relation $ic1$ in the 9^+-intersection calculus module whereas the connection $\mathsf{sim}_2^1(ic1, g1)$ relates a spatial relation $ic1$ in the 9^+-intersection calculus module to a spatial modality instance $g1$ in the *GUM-Space* module. Both connections can be assigned values for higher or lower similarities. In addition, similarity relations within the 9^+-intersection calculus module are illustrated. These examples indicate direct neighbors in the neighborhood graph of the calculus, for instance, $ic2$, $ic3$, and $ic4$ are direct neighbors of $ic1$. This example points out that closely related calculus relations are not necessarily closely related to their linguistic counterparts. Although $ic2$ and $ic4$ are possible situational interpretations of $g2$, $ic3$ is less likely to be an interpretation of $g2$, as the linguistic expression "X is going out of Y" implies that the end point of the directed line is outside the region.

Technically, the similarity can be modeled as a 'heterogeneous similarity' of objects drawn from conceptually different domains on the basis of distance functions. This means that similarities that are defined between different categories or relations can be metrically compared or combined, i.e., calculation rules for similarities are available such as in distance logics Hois and Kutz [2008a]. In particular, we can define a similarity space consisting of a set of similarities and their similarity measure. To specify this space for different modules, a *heterogeneous similarity space* $\mathbb{H} = \langle \mathbb{S}_1, \mathbb{S}_2, f_1^2, f_2^1 \rangle$ is required that consists of, for $i = 1, 2$, the similarity spaces $\mathbb{S}_i = \langle \mathcal{S}_i, f_i \rangle$, and heterogeneous similarity measures $f_i^{\bar{i}} : \mathcal{S}_i \times \mathcal{S}_{\bar{i}} \mapsto \mathbb{R}_{0,\infty}^+$.

Between these spaces, similarity relations (counterparts) can be defined, i.e., $b_{\bar{i}} \in \mathcal{S}_{\bar{i}}$ is an $\bar{i}$-counterpart of $a_i \in \mathcal{S}_i$ if $f_i^{\bar{i}}(a_i, b_{\bar{i}}) = \inf\{f_i^{\bar{i}}(a_i, b) \mid b \in \mathcal{S}_{\bar{i}}\} < \infty$. This is written as $\mathsf{Cp}_i^{\bar{i}}(a_i, b_{\bar{i}})$ in short. Two relations can now be defined: $\mathsf{Cp}_i^{\bar{i}} \subseteq \mathcal{S}_i \times \mathcal{S}_{\bar{i}}, i = 1, 2$. These provide a way to define similarity relations between categories or relations from different or even the same ontology modules. However, such counterparts may or may not be unique and applications may require this to select amongst the elements with maximal similarity a unique element, e.g., according to some external criteria. A way to make counterparts unique is thus to use a choice function. Note also that $b_{\bar{i}}$ may be an $\bar{i}$-counterpart of a_i without a_i being an i-counterpart of $b_{\bar{i}}$. Hence, one category can be similar to another but not vice versa, unless the relation is defined to be bi-directional and unique.

In summary, the scenarios for connecting spatial language with qualitative space have shown methods for spatial language interpretation. Ontological formalisms and empirical results were presented and discussed. They have primarily demonstrated that the motivating distinction between linguistic and qualitative spatial modules, and its respective specification by modules complying with the spatial perspectives, leads to a flexible and robust interplay between both modules. The distinction also allows us to formally structure and present spatial information based on available linguistic characteristics and to deal with aspects of ambiguity and uncertainty caused by linguistic underspecifications, granularity differences, context dependencies, metaphors, etc. Furthermore, the mapping from linguistic data to qualitative spatial relations may lead to more efficient and effective reasoning strategies, e.g., the mapping from spatial language to a qualitative formal model allows spatial reasoning with this model, whereas spatial reasoning with spatial language itself is hardly feasible [Bateman et al., 2010b].

In contrast to our approach to map language to space, some spatial models exist that define spatial relations on the basis of linguistic terms (e.g., Kurata and Shi [2008]). Such models reflect, however, not spatial language semantics but specific meanings that are determined by the model's axioms, i.e., these models are typically fine-tuned for computational requirements. The diverse use of language requires yet more complex and flexible logical approaches [Bateman, 2010a]. Such approaches need to take into account contributing aspects for spatial language interpretations, such as context or world knowledge, in order to achieve a mapping from spatial language to qualitative representations, as we have discussed above, and for which we have provided modular formalisms and empirical evaluations.

6.5 Chapter Summary

In this chapter, three application scenarios were presented that use the spatial ontology modules and their extensions from the previous chapters. The applications showed the use of spatial modules within the distinct areas of architectural design/assisted living, visual recognition, and spatial language interpretation. All scenarios highlighted that several spatial perspectives are needed to fully model their respective domains. The ontology modules complying with different spatial perspectives (chapter 4) together with their combinations and uncertainty integrations (chapter 5) were able to support the application goals and to achieve the application tasks.

The use of the spatial ontology modules demonstrated their broad applicability, reusability, and generalization. As some spatial modules have been used in more than one application scenario, it has been demonstrated that it is feasible to apply these modules in other systems and applications as well. Some ontology modules may therefore need to be adjusted to satisfy the requirements of a particular application. However, they can build on the spatial perspective distinction and be refined or extended accordingly, as shown in this chapter.

7

Conclusions and Outlook

The goal of this thesis was to analyze and categorize the diverse types of spatial information and to provide ontology modules for them. For this purpose, we have applied modularity techniques and, wherever necessary, extended them to achieve the intended spatial specifications. Additionally, the resulting ontology modules were evaluated in different application scenarios of spatial assistance systems. We shortly summarize results and outcomes of this work with regard to these initial goals and we discuss related and future work in this chapter.

7.1 Summary

Chapter 3 shows that a large variety of different representations for spatial information has been specified in the research literature so far. Here, we identified quantitative/qualitative, abstract, domain-specific, and multimodal types of spatial information, also called spatial perspectives. These perspectives differ with regard to the ontological categorizations they can provide. They determine the characteristics of their corresponding ontological modules, in particular, definitions of categories, detail of axiomatization, grade of generalization, level of granularity, semantic motivation, and design aspects. Hence, spatial perspectives support the specification of spatial ontology modules and indicate their structural distinctions. Users or ontology designers can thus use the perspectives as a general classification of spatial information types. Combinations of modules from different perspectives allow a comprehensive, clearly structured, and modularly designed spatial representation.

Chapter 4 introduce several spatial ontology modules that have been developed, refined, or integrated in this thesis according to the distinction of the spatial perspectives. Primarily motivated by the applicability for specific spatial systems or tasks, the presented ontology modules are designed for structural building information and region-based data to support architectural design applications, for visually perceivable spatial entities to support vision recognition, for spatial actions and home automation to support spatial assistance systems, and for linguistic features to support spatial language tasks. Each module was developed according to formal ontology design principles, namely by making categories and relations of their domain as closely as possible to reality, e.g., by using external data or existing resources that are available for the domain. Requirements, purposes, and aims were also defined for each module, and the results have been evaluated in terms of applicability, performance, or cognitive adequacy.

In addition to developing individual ontology modules, their combinations are often required to exchange information among perspectives, e.g., to achieve certain spatial tasks. Chapter 5 demonstrates how the different spatial ontology modules were combined with each other by using modularity techniques. Specifically the methods extension and refinement, matching, and connection were used for combining spatial ontology modules: extension/refinement integrate contents of modules, matching relates modules to external data formats, and connections provide links between modules. Based on the perspectives of the modules and their intended meanings, combination methods have been chosen for individual spatial modules. Some of the modules are combined with uncertain information to further assist spatial systems by specialized ontology modules. As a consequence, these spatial modules enable spatial systems to extend the ontological representations with beliefs about instances or to combine the ontological representations with statistical data about the domain.

The distinction of spatial information and the use of the spatial ontology modules and their extensions is shown in chapter 6, where several spatial assistance systems demonstrate how combined spatial ontology modules can accomplish their designated tasks and purposes. Architectural design and assisted living applications have shown that the spatial perspective distinction gives a clear structure to the domain and its modular combination can directly support application-specific functions. Visual recognition applications have shown that the combination of ontological representations and

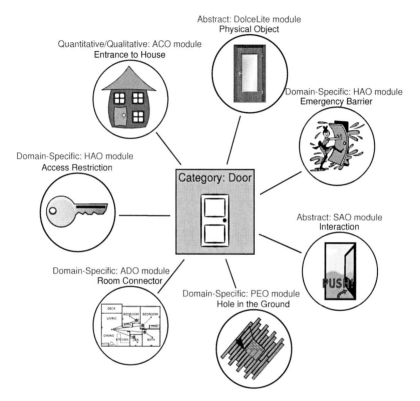

Figure 7.1: Spatial Perspectives on the Category Door. Several ontology modules describing characteristics of a door and the spatial perspectives can be determined by their spatial perspectives. Every thematic interpretation of a door can now be related to its spatial perspective and to the spatial module in which it can be defined (see chapter 4 for explanations of the individual modules). (Images taken from http://openclipart.org, visited on November 15, 2011)

uncertain information can support the identification of visually represented spatial scenes. Spatial language interpretation applications have shown that the combination of rather distinct spatial information and the connection with statistical data provide a solution to interpret linguistically expressed spatial situations.

Overall, this thesis shows that spatial information can be seen and described from different perspectives, such as spatial calculi, language, visual features, terminological classifications, quantitative data, or task-specific aspects. Ontological modules can specify these distinct perspectives, and their extensions can be used to combine individual modules. This not only yields a clear representation of spatial information but it also directly supports application-specific tasks and requirements of various spatial assistance systems. We can now also solve the complex example of the different characteristics that a category Door may have (cf. figure 3.1) by analyzing and specifying the different perspectives on it, as shown in figure 7.1.

7.2 Discussion

As pointed out in the introduction: if a system represents spatial information, it needs to define which aspects of space should be represented and how these aspects need to be specified. The spatial perspectives presented in this thesis provide the system with a general classification of the spatial types it may need to represent. Based on these perspectives, ontology modules can be developed or reused for the system's purpose. The scope of those ontology modules is then determined by the perspectives they comply with, such as axiomatization, granularity, or design aspects. For example, if a system is supposed to represent architectural types (floors, rooms, bridges, tunnels), it primarily needs to model domain-specific perspectives and if necessary qualitative/quantitative or abstract ones as well. Specific ontology modules for such architectural entities can then be reused or designed according to the spatial perspective in a modular way. This can result in the development of different ontology modules with different spatial perspectives, specifications and characteristics. Their combinations can be specified by means of their perspectives and potential uncertainties can be integrated. This way, a detailed and structured ontological representation becomes available for the system. Hence, the modular structure developed in this thesis provides a representation of the thematically different aspects of spatial information, it allows the use of application-specific ontologies where necessary, it reduces complexity in terms of the represented spatial knowledge, and accordingly enhances reasoning practicability.

The applications in chapter 6 have shown that spatial ontologies and their differentiation based on the spatial perspectives can indeed clarify the system's domain and support the system's tasks and purposes. For example, the architectural design application has illustrated that a variety of different aspects of spatial information contributes to the overall representation of courthouses (section 6.2.1). High-level architectural design requirements can be specified adequately only if these different types are disentangled and made explicit, e.g., by the modular structure of the ontological representation. The spatial perspectives allow the system to make the different types explicit. They also determine combinations of the different spatial types.

As uncertainty is typically inherent in real world applications, spatial ontological representations need to deal with such uncertain information. An example has been presented by the linguistic application (section 6.4.2), where empirical data can be used to learn link relations between two different ontology modules. As empirical data is limited and can be noisy, the resulting link relations have different grades of accuracy. We have thus discussed a method to represent levels of accuracy by using similarities. This example also shows that the ontological modules that are based on the spatial perspectives are flexible enough to integrate extensions, e.g., link relations that can be specified as similarities between modules. However, a concrete spatial system still needs to deal with such specific extensions and implement them individually, whenever necessary.

However, exhaustiveness of the spatial perspectives and their ontological modules cannot be fully guaranteed. Even though we have applied the developed ontology modules in several application scenarios and we have taken into account a large number of existing approaches and different disciplines, the perspectives cannot assure completeness simply due to the open world criterion. Nevertheless, a general spatial structure is provided by these perspectives. They also guide the development of new ontology modules and their characteristics. Although they were sufficient for the different spatial application examples in which they were used, future developments may discover potential limits.

The general advantages of using modularly designed ontologies, namely reusability, interoperability, and decentralization, were clearly and once again confirmed by the results in this thesis. For example, the region-based ontology module *PEO* has been reused several times (reusability). It has also been combined with other modules by refinement and connection (interoperability). Its design and specification is only constrained by its own module requirements (decentralization). Furthermore, the *PEO* module refinement has demonstrated that the reuse of the *PEO* module in a new application, namely a different vision recognition system, can be achieved successfully (section 6.3.2).

Even though the developed ontology modules have been evaluated well with regard to their purpose and application, details in their specification might be challenged. For example, the *HAO* module contains complex category definitions that determine the condition of an assisted living environment (section 4.2.3.3). As the *HAO* module was primarily designed to specify valid situations or behaviors of an assisted living environment, it seems evident that rules provided by the Semantic Web Rule Language (SWRL [Horrocks et al., 2004]) might be more practical for such specifications. Although this language has technically been integrated into Protégé and is equipped with respective reasoning mechanisms, the spatial assistance behavior that has been modeled in the *HAO* module simply did not require the expressiveness of rules in SWRL. Furthermore, Gasse et al. [2008] have proven that SWRL rules can be rewritten in standard $\mathcal{SROIQ}$ ontologies entirely. They provide a process that translates rules into $\mathcal{SROIQ}$ constraints by removing inverses, reformulating rules as role inclusions, and (optionally) adding the .Self operator for referencing. Hence, as rules were not necessary to model the spatial environmental situations and as they can be rewritten in OWL, we did not use SWRL when developing the spatial ontology modules.

Technically, we have often used description logics and OWL for specifying the spatial ontology modules and their combinations. $\mathcal{E}$-connections [Kutz et al., 2004] were used as a connection method. However, approaches have been developed that allow other types of mappings between ontologies. The semantics of these mappings are defined, for instance, by distributed description logics [Borgida and Serafini, 2003] or interface-based modules [Ensan, 2008]. These methods range from single mappings of two instances from different ontologies to complex formulae that describe relations between several categories from different ontologies. Regardless of the complexity of these approaches, parts of one ontology are related to parts of another as defined by certain semantics. Hence, such approaches could have also been used in this thesis. However, their applicability and reasoning support in the spatial application examples, as presented here, and their suitability with regard to the spatial perspectives remain to be investigated. Regardless of combination approaches, all spatial ontologies could have been technically composed as one monolithic ontology comprising all different perspectives, i.e., a union of all modules. However, this would increase complexity and limit reasoning practicability. It can also cause inconsistencies (cf. section 3.1) that need to be fixed, or such inconsistencies remain undetected if reasoning over the monolithic ontology is too complex. The flexibility of using single modules required by an application as well as the clearly separated spatial types of information by the perspectives would be lost. Furthermore, a conglomerate of all perspectives on space is hardly ever required by a spatial system in a specific domain.

Note, that some of the combinations of spatial modules result in an ontological representation that is similar with a monolithic representation. For example, the *PEO* module inherites and refines the DOLCE-Lite module, which results in one representation (section 5.1.1), and accordingly entails all specifications from DOLCE-Lite categories and relations. However, we have seen that the *PEO* module has refined primarily three categories from DOLCE-Lite. Hence, if a spatial system uses the *PEO* module, it can focus on those refined categories, e.g., by analyzing their *PEO* namespace. Further inherited information with different namespaces can be avoided by the system, as long as the ontology module stays consistent. The system can thus ignore details of other entailed modules if they are not specified directly in the *PEO* module. Even though some formalisms provide a hiding mechanism, e.g., CASL [Mossakowski et al., 2008], to conceal information from inherited modules, the logical assumptions of those hidden parts still have to be satisfied in the inheriting module. Here, the hidden categories and relations are also implicitly part of the inheriting module.

Another advantage of the module structure is the interchangeability of spatial modules. As we have seen in the linguistic example (section 6.4), spatial linguistic terms can be interpreted in terms of spatial calculi. In particular, a linguistic module was connected with modules for qualitative spatial representations, namely to a module for the 9^+-intersection calculus, to a module for the double-cross calculus, and to a module for the region connection calculus. Three times, the module with a linguistic perspective was combined with a module with a qualitative perspective by using link relations in an identical way. This identical combination is directly supported by the spatial perspectives (section 5.1), which generalizes combinations across modules and which supports the interchangeability of spatial modules, as expected from modularity in general.

Hence, with the spatial perspectives presented in this thesis, complex domains of space can be specified by ontology modules adequately. An example of such a complex spatial domain is the area of urban planning. The well-known contribution by Lynch [1960] has shown several key aspects that all cities have in common, in which the diversity and the different spatial perspectives on the same domain are reflected. Lynch lists paths, edges, districts, nodes, and landmarks as 'physical elements' that are common in each city structure. Paths (abstract) provide a way to move (action) across the city (domain-specific), edges (abstract) are boundaries (abstract) between regions (qualitative) and possibly barriers (domain-specific), districts (domain-specific) are larger two-dimensional regions (quantitative/qualitative) that have an identifying character (domain-specific), nodes (abstract) are point-like strategic spots or concentrations (abstract) with high importance that can be entered (abstract), and landmarks (abstract) are point-references (abstract) which cannot be entered but which serve as reference points (abstract). Several modules from different perspectives are thus involved in these definitions. In principle, ontology modules would also need to take into account social behavior and crowd movements in certain environments, in particular with regard to cultural differences. However, this goes beyond the scope of spatial information as addressed in this thesis.

New spatial ontology modules can be developed making use of the spatial perspectives in order to gain a clear separation between different spatial contents and ontological specifications. Existing spatial ontologies can be classified with regard to the perspectives and if necessary restructured in terms of the perspectives into spatial modules. The perspectives thus provide a guideline for design criteria of spatial ontologies. Spatial systems are then able to select along the perspectives their required types of spatial information for a certain task or purpose. The spatial perspectives determine how different spatial modules can be thematically combined with each other by using different technical strategies. The spatial applications in this thesis have also demonstrated how uncertain information can be related with the spatial ontology modules.

7.3 Outlook

As space is inherent in many domains, our application and evaluation scenarios were able to analyze only a few of them. Many other spatial topics and tasks can be represented by using modular ontologies complying with the spatial perspectives. For instance, space in the sense of outer space has not been addressed in this thesis. Also the huge research area of geographical information systems and geographic ontology modules has been noted only briefly. Even in the applications we have demonstrated not all topics that can be modeled and have been investigated. In the architectural domain, commercial and industrial buildings have not been addressed so far. They may have further unique characteristics that can be designed by modules from the different spatial perspectives. Similarly, architecturally entirely different constructions, such as cruise liners (especially together with tourism-related ontologies), were not discussed.

Related work that is associated with the topics presented in this thesis and that could be integrated with the spatial perspectives is an ontology of measurements: such an ontology aims at grounding spatial types of information [Probst, 2006], which is thus an aspect of the quantitative spatial perspective. Other related approaches are part of the research area concerned with route instructions. Such approaches could also be integrated more closely with regard to several spatial perspectives. The approach presented in MacMahon et al. [2006], for instance, also distinguishes among linguistic, domain-specific, and action-related types of spatial instructions, which may comply directly with the spatial perspectives shown in this thesis.

On a broader scale, the spatial perspectives could be extended to more types of information than just spatial. As a first step, temporal information could be related with the spatial perspectives as well, to achieve a full spatial-temporal representation. Eventive categories could then be anchored in space. Such representations have, for example, been investigated in Bennett and Galton [2004], where temporal categories and characteristics have been identified. In further steps, special non-spatial domains could be analyzed and categorized by domain-specific perspectives that determine possible ontology modules for the domains. Such thematic non-spatial perspectives could comprise social, emotional, or political types of information.

Other technical directions can also be pursued in future work. For example, other ontology languages that support the specification of modules can be used for designing the spatial modules. An example is Common Logic, a family of knowledge representation languages, for which semantics of ontology modules have been proposed [Neuhaus and Hayes, 2012]. However, all ontology modules would need to be translated to this language in order to use it, which adds additional translation effort and reduces reasoning practicability. Specifically in the area of assisted living environments, ontology specifications in Prolog have been presented and applied [Normann et al., 2009], which could also be investigated further. Another example is to use the Distributed Ontology Language, a standardized framework for ontologies that are specified in heterogeneous logics [Mossakowski et al., 2012]. Here, several ontology languages are supported and also their links between each other can be specified. This framework is furthermore scheduled to become an ISO standard in 2015.

Another technical question, that has not been addressed in this thesis, is the use and definition of 'primary' instances. A spatial system can use different spatial ontology modules for representing its multiperspectival domain. An object in reality can thus have different instantiations in the different ontology modules. The question could arise if it is necessary to define one of these instances as the 'primary' one, i.e., as the leading instance. For example, if the properties of the primary instance are changed, connected instances in other spatial modules might have to be changed automatically. Although we did not need such primary instances (i.e., we treat all instances across modules equally), it remains an open issue if primary instances are required, e.g., if applications access only one of the spatial modules. Such primary instances, however, would have to be defined in a specific way that prevents multiple primary instances of the same entity in reality.

Another open question concerns tool support that could be provided in the future to select spatial perspectives and their modules. This could also comprise visual aids and extended meta-information about the developed spatial ontology modules. In other approaches, this is supported by open repositories, e.g., similar to the BioPortal ontology repository for biomedical ontologies[1]. Such approaches have recently been researched in more detail by the Open Ontology Repository (OOR) Initiative[2]. Here,

---

[1] http://bioportal.bioontology.org (visited on March 09, 2012)

[2] http://ontolog.cim3.net/cgi-bin/wiki.pl?OpenOntologyRepository (visited on March 09, 2012)

technical requirements, metadata, and standardizations are investigated and proposed to provide controlled access to ontological specifications.

Extending spatial ontology modules with types of uncertainties beyond the demonstrations in this thesis is also open for future research. In particular, transformations between different uncertainty theories are still an open issue. These transformations provide an approximation for translating one uncertainty theory into another [Umkehrer and Schill, 1995]. This way, it might be less important which specific theories for uncertainties has been used in an ontological representation. Concrete uncertainty values might be translated between uncertainty theories. However, related reasoning techniques as indicated by the different uncertainty theories will change as well. Specifically for vision recognition, empirical data could also be analyzed with regard to correlations between object types and their spatial relations. These correlations may yield statistical likelihood that can be used for image classifications. A similar approach has been proposed in Hudelot et al. [2008], where spatial relations have been used for image classifications with fuzzy set theory.

Another issue that has not been addressed thoroughly in this thesis is scalability of the developed ontology modules. For example, the *HAO* module reflects some of the conditions of an assisted living environment. However, large-scale systems, e.g., the BAALL assisted living environment [Krieg-Brückner et al., 2009], may need to extend the ontology modules in several ways to represent their internal facilities and behaviors. Here, time performance and complexity issues may arise. These aspects are also relevant in dialogue systems, when spatial language interpretations as presented in this thesis need to be processed further. As time-critical applications were not part of the scope of this thesis, this is left for future work.

Certainly there may be even more directions that can be pursued in the future with regard to the topics in this thesis. Yet, these directions can build on the results presented here. We have shown that ontological specifications provide a representation formalism that focuses on relevance, clarity, consistency, and compatibility. In addition, modularity even improves reusability, interchangeability, and diversity. We have specifically demonstrated how these aspects can be applied in the spatial domain and with spatial perspectives. A series of application use cases has been evaluated and show how the spatial perspectives work to reflect spatial information.

In brief, the aim of this thesis was to show that the diversity of spatial types of information cannot be modeled adequately without taking into account spatial perspectives, and we have presented and discussed solutions for such a perspectival, ontological, modular representation.

"And where is a bird which is 'on the back' of a buffalo?

That's right: On 'top' of it." [Trask and Mayblin, 2000, p. 166]

Appendix A

Ontology Appendix

Attached versions of the ontology modules presented in chapter 4 are available at
`http://www.informatik.uni-bremen.de/~joana/ontology/SpatialOntologies.html`

References

AIELLO, MARCO; PRATT-HARTMANN, IAN E.; and VAN BENTHEM, JOHAN (2007). What is Spatial Logic? In: Marco Aiello; Ian E. Pratt-Hartmann; and Johan van Benthem (eds.), Handbook of Spatial Logics, pp. 1–11. Springer.

ALANI, HARITH and BREWSTER, CHRISTOPHER (2006). Metrics for ranking ontologies. In: Workshop on Evaluating Ontologies for the Web (EON'06).

ALLEN, JAMES F. and FRISCH, ALAN M. (1982). What's in a semantic network? In: 20th Annual Meeting on Association for Computational Linguistics, pp. 19–27. Association for Computational Linguistics.

ANDERSON, ANNE H.; BADER, MILES; BARD, ELLEN G.; BOYLE, ELIZABETH; DOHERTY, GWYNETH; GARROD, SIMON; ISARD, STEPHEN; KOWTKO, JACQUELINE; McALLISTER, JAN; MILLER, JIM; SOTILLO, CATHERINE; THOMPSON, HENRY; and WEINERT, REGINA (1991). The HCRC Map task Corpus. Language and Speech, 34(4), 351–366.

ARAÚJO, RUDI and PINTO, HELENA SOFIA (2008). Towards Semantics-based Ontology Similarity. In: Pavel Shvaiko; Jérôme Euzenat; Fausto Giunchiglia; and Bin He (eds.), 2nd International Workshop on Ontology Matching (OM'07). CEUR Workshop Proceedings.

ATENCIA, MANUEL and SCHORLEMMER, MARCO (2007). A Formal Model for Situated Semantic Alignment. In: 6th International Joint Conference on Autonomous Agents and Multi-Agent Systems (AAMAS'07), pp. 1270–1277. IFAMAS.

AUER, PETER; BILLARD, AUDE; BISCHOF, HORST; BLOCH, ISABELLE; BOETTCHER, PIA; BÜLTHOFF, HEINRICH; BUXTON, HILARY; CHRISTENSEN, HENRIK; COHN, TONY; COURTNEY, PATRICK; CROOKELL, ANDREW; CROWLEY, JAMES; DICKINSON, SVEN; EBERST, CHRISTOF; EKLUNDH, JAN-OLOF; FISHER, BOB; FÖRSTNER, WOLFGANG; GILBY, JOHN; GRANLUND, GOESTA; HLAVAC, VACLAV; KITTLER, JOSEF; KROPATSCH, WALTER; LEONARDIS, ALES; LITTLE, JIM; METTA, GIORGIO; NAGEL, HANS-HELLMUT; NEBEL, BERNHARD; NEUMANN, BERND; NIEMANN, HEINRICH; PALETTA, LUCAS; PINZ, AXEL; PIRRI, FIORA; SAGERER, GERHARD; SANDINI, GIULIO; SCHIELE, BERNT; SIMPSON, REBECCA; SOMMER, GERALD; TSOTSOS, JOHN; THONNAT, MONIQUE; VERNON, DAVID; and

VINCZE, MARKUS (2005). A Research Roadmap of Cognitive Vision. Tech. rep., IST Project IST-2001-35454. Version 5.0.

BAADER, FRANZ; CALVANESE, DIEGO; MCGUINNESS, DEBORAH; NARDI, DANIELE; and PATEL-SCHNEIDER, PETER (2003). The Description Logic Handbook. Cambridge University Press.

BAADER, FRANZ; HORROCKS, IAN; and SATTLER, ULRIKE (2004). Description Logics. In: Steffen Staab and Rudi Studer (eds.), Handbook on Ontologies, pp. 3–28. Springer.

BAADER, FRANZ; HORROCKS, IAN; and SATTLER, ULRIKE (2007). Description Logics. In: Frank van Harmelen; Vladimir Lifschitz; and Bruce Porter (eds.), Handbook of Knowledge Representation. Elsevier.

BAJCSY, RUZENA (1988). Active Perception. In: Proceedings of the IEEE, vol. 76, pp. 996–1005.

BALDWIN, CARLISS Y. and CLARK, KIM B. (2006). Modularity in the Design of Complex Engineering Systems. In: Ali Minai; Dan Braha; and Yaneer Bar Yam (eds.), Complex Engineered Systems: Science Meets Technology. Springer.

BARCLAY, MICHAEL and GALTON, ANTONY P. (2008). An Influence Model for Reference Object Selection in Spatially Locative Phrases. In: Christian Freksa; Nora S. Newcombe; Peter Gärdenfors; and Stefan Wölfl (eds.), Spatial Cognition VI – Learning, Reasoning, and Talking about Space, pp. 216–232. Springer.

BATEMAN, JOHN (2010a). Situating spatial language and the role of ontology: issues and outlook. Linguistics and Language Compass, 4(8), 639–664.

BATEMAN, JOHN; BORGO, STEFANO; LÜTTICH, KLAUS; MASOLO, CLAUDIO; and MOSSAKOWSKI, TILL (2007a). Ontological Modularity and Spatial Diversity. Spatial Cognition and Computation, 7(1), 97–128.

BATEMAN, JOHN; COHN, ANTHONY; and PUSTEJOVSKY, JAMES (2010a). Spatial Representation and Reasoning in Language: Ontologies and Logics of Space. Dagstuhl Seminar 10131 Proceedings.

BATEMAN, JOHN; HOIS, JOANA; and FARRAR, SCOTT (2009). Spatial Ontology Baseline — Project OntoSpace: Deliverable D2 [2nd edition]. Tech. rep., Spatial Cognition Research Center SFB/TR8, University of Bremen.

BATEMAN, JOHN; TENBRINK, THORA; and FARRAR, SCOTT (2007b). The Role of Conceptual and Linguistic Ontologies in Discourse. Discourse Processes, 44(3), 175–213.

BATEMAN, JOHN A. (2010b). Language and Space: a two-level semantic approach based on principles of ontological engineering. International Journal of Speech Technology, 13(1), 29–48.

BATEMAN, JOHN A.; HENSCHEL, RENATE; and RINALDI, FABIO (1995). Generalized Upper Model 2.0: Documentation. Tech. rep., GMD/Institut für Integrierte Publikations- und Informationssysteme, Darmstadt, Germany.

BATEMAN, JOHN A.; HOIS, JOANA; ROSS, ROBERT; and TENBRINK, THORA (2010b). A linguistic ontology of space for natural language processing. Artificial Intelligence, 174(14), 1027–1071.

BELLENGER, AMANDINE and GATEPAILLE, SYLVAIN (2010). Uncertainty in Ontologies: Dempster-Shafer Theory for Data Fusion Applications. In: Workshop on the Theory of Belief Functions.

BENNETT, BRANDON (1996). Modal logics for qualitative spatial reasoning. Journal of the Interest Group on Pure and Applied Logic, 4, 23–45.

BENNETT, BRANDON (2005). Modes of concept definition and varieties of vagueness. Applied Ontology, 1(1), 17–26.

BENNETT, BRANDON and AGARWAL, PRAGYA (2007). Semantic categories underlying the meaning of 'Place'. In: Stephan Winter; Matt Duckham; Lars Kulik; and Benjamin Kuipers (eds.), Spatial Information Theory: Proceedings of the 8th International Conference (COSIT'07), pp. 78–95. Springer.

BENNETT, BRANDON and GALTON, ANTONY P. (2004). A unifying semantics for time and events. Artificial Intelligence, 153(1-2), 13–48.

BENNETT, BRANDON; MAGEE, DEREK R.; COHN, ANTHONY G.; and HOGG, DAVID C. (2008). Enhanced tracking and recognition of moving objects by reasoning about spatio-temporal continuity. Image and Vision Computing, 26(1), 67–81. Cognitive Vision-Special Issue.

BERNERS-LEE, TIM; HENDLER, JAMES; and LASSILA, ORA (2001). The Semantic Web. Scientific American Magazine. Retrieved 2008-03-26.

BHATNAGAR, RAJ K. and KANAL, LAVEEN N. (1986). Handling Uncertain Information: A Review of Numeric and Non-Numeric Methods. In: Laveen N. Kanal and John F. Lemmer (eds.), Uncertainty in Artificial Intelligence, pp. 3–26. North-Holland.

BHATT, MEHUL and DYLLA, FRANK (2009). A Qualitative Model of Dynamic Scene Analysis and Interpretation in Ambient Intelligence Systems. International Journal of Robotics and Automation, 24(3).

BHATT, MEHUL; DYLLA, FRANK; and HOIS, JOANA (2009). Spatio-Terminological Inference for the Design of Ambient Environments. In: Kathleen Stewart Hornsby; Christophe Claramunt; Michel Denis; and Gérard Ligozat (eds.), 9th International Conference on Spatial Information Theory (COSIT'09), pp. 371–391. Springer.

BHATT, MEHUL; HOIS, JOANA; and KUTZ, OLIVER (2012). Modelling Form and Function in Architectural Design. Applied Ontology, 7(3), 233–267.

BOBILLO, FERNANDO and STRACCIA, UMBERTO (2009). An OWL Ontology for Fuzzy OWL 2. In: Jan Rauch; Zbigniew Ras; Petr Berka; and Tapio Elomaa (eds.), Foundations of Intelligent Systems, pp. 151–160. Springer.

BORGIDA, ALEX and SERAFINI, LUCIANO (2003). Distributed Description Logics: Assimilating Information from Peer Sources. Data Semantics, pp. 153–184.

BORGO, STEFANO; GUARINO, NICOLA; and MASOLO, CLAUDIO (1996). A pointless theory of space based on strong congruence and connection. In: Luigia Carlucci Aiello; Jon Doyle; and Stuart C. Shapiro (eds.), 5th International Conference on Principles of Knowledge Representation and Reasoning (KR'96), pp. 220–229. Morgan Kaufmann.

BOUQUET, PAOLO; GIUNCHIGLIA, FAUSTO; VAN HARMELEN, FRANK; SERAFINI, LUCIANO; and STUCKENSCHMIDT, HEINER (2004). Contextualizing ontologies. Journal of Web Semantics, 1(4), 325–343.

BRACHMAN, RONALD J. and LEVESQUE, HECTOR J. (2004). Knowledge representation and reasoning. Morgan Kaufmann.

BRACHMAN, RONALD J. and SCHMOLZE, JAMES G. (1985). An Overview of the KL-ONE Knowledge Representation System. Cognitive Science, 9(2), 171–217.

BRANK, JANEZ; GROBELNIK, MARKO; and MLADENIĆ, DUNJA (2005). A survey of ontology evaluation techniques. In: Conference on Data Mining and Data Warehouses (SiKDD'05).

BREWSTER, CHRISTOPHER; ALANI, HARITH; DASMAHAPATRA, SRINANDAN; and WILKS, YORICK (2004). Data-driven ontology evaluation. In: Proceedings of the 4th International Conference on Language Resources and Evaluation. European Language Resources Association.

BROOKS, RODNEY A. (1999). Intelligence without Representation. In: The Early History of the New AI, chap. 5, pp. 79–101. A Bradford Book, MIT Press.

BUCHANAN, BRUCE G. and SHORTLIFFE, EDWARD H. (eds.) (1984). Rule-Based Expert Systems: The MYCIN Experiments of the Stanford Heuristic Programming Project. Addison-Wesley Publishing Company.

BUCHE, PATRICE; DIBIE-BARTHÉLEMY, JULIETTE; and IBANESCU, LILIANA (2008). Ontology Mapping Using Fuzzy Conceptual Graphs and Rules. In: Peter W. Eklund and Ollivier Haemmerlé (eds.), 16th International Conference on Conceptual Structures (ICCS'08), pp. 17–24. CEUR Workshop Proceedings.

BURTON-JONES, ANDREW; STOREY, VEDA C.; SUGUMARAN, VIJAYAN; and AHLUWALIA, PUNIT (2005). A semiotic metrics suite for assessing the quality of ontologies. Data and Knowledge Engineering, 55(1), 84–102.

CALEGARI, SILVIA and CIUCCI, DAVIDE (2007). Fuzzy Ontology and Fuzzy-OWL in the KAON Project. In: IEEE International Conference on Fuzzy Systems (FUZZ-IEEE'07), pp. 1415–1420. IEEE Computer Society.

CALÍ, ANDREA; LUKASIEWICZ, THOMAS; PREDOIU, LIVIA; and STUCKENSCHMIDT, HEINER (2007). A Framework for Representing Ontology Mappings under Probabilities and Inconsistency. In: Fernando Bobillo; Paulo Cesar G. da Costa; Claudia d'Amato; Nicola Fanizzi; Francis Fung; Thomas Lukasiewicz; Trevor Martin; Matthias Nickles; Yun Peng; Michael Pool; Pavel Smrž; and Peter Vojtás (eds.), Third Workshop on Uncertainty Reasoning for the Semantic Web (URSW'07). CEUR Workshop Proceedings.

CARLSON, LAURA A. and VAN DER ZEE, EMILE (eds.) (2005). Functional features in language and space: Insights from perception, categorization and development. Oxford University Press.

CASTRO, ALEXANDER GARCIA; ROCCA-SERRA, PHILIPPE; STEVENS, ROBERT; TAYLOR, CHRIS; NASHAR, KARIM; RAGAN, MARK A; and SANSONE, SUSANNA-ASSUNTA (2006). The use of concept maps during knowledge elicitation in ontology development processes – the nutrigenomics use case. BMC Bioinformatics, 7(267).

CHANDRASEKARAN, BALAKRISHNAN; JOSEPHSON, JOHN R.; and BENJAMINS, V. RICHARD (1999). What Are Ontologies, and Why Do We Need Them? IEEE Intelligent Systems, 14(1), 20–26.

CHUA, SOOK-LING; MARSLAND, STEPHEN; and GUESGEN, HANS (2009). Spatio-Temporal and Context Reasoning in Smart Homes. In: Mehul Bhatt and Hans W. Guesgen (eds.), Workshop on Spatial and Temporal Reasoning for Ambient Intelligence Systems (STAMI'09), pp. 9–20.

CIMIANO, PHILIPP (2006). Ontology Learning and Population from Text: Algorithms, Evaluation and Applications. Springer.

CLARKE, BOWMAN L. (1981). A calculus of individuals based on 'connection'. Notre Dame Journal of Formal Logic, 22(3), 204–218.

COHN, ANTHONY G.; BENNETT, BRANDON; GOODAY, JOHN; and GOTTS, NICHOLAS M. (1997a). Qualitative Spatial Representation and Reasoning with the Region Connection Calculus. GeoInformatica, 1, 275–316.

COHN, ANTHONY G.; BENNETT, BRANDON; GOODAY, JOHN; and GOTTS, NICHOLAS M. (1997b). Representing and Reasoning with Qualitative Spatial Relations. In: Oliviero Stock (ed.), Spatial and Temporal Reasoning, pp. 97–132. Kluwer Academic Publishers.

COHN, ANTHONY G. and HAZARIKA, SHYAMANTA M. (2001). Qualitative Spatial Representation and Reasoning: An Overview. Fundamenta Informaticae, 43, 2–32.

COMMON LOGIC WORKING GROUP (2003). Common Logic: Abstract syntax and semantics. Tech. rep.

CONDOTTA, JEAN-FRANCOIS; SAADE, MAHMOUD; and LIGOZAT, GÉRARD (2006). A Generic Toolkit for n-ary Qualitative Temporal and Spatial Calculi. In: 13th International Symposium on Temporal Representation and Reasoning (TIME'06), pp. 78–86. IEEE Computer Society.

COSTA, PAULO C. G. and LASKEY, KATHRYN B. (2006). PR-OWL: A framework for probabilistic ontologies. In: Brandon Bennett and Christiane Fellbaum (eds.), Formal Ontology in Information Systems (FOIS'06), pp. 237–249. IOS Press.

COSTA, PAULO C. G.; LASKEY, KATHRYN B.; and LASKEY, KENNETH J. (2008). PR-OWL: A Bayesian Ontology Language for the Semantic Web. In: Paulo C. G. Costa; Claudia D'Amato; Nicola Fanizzi; Kathryn B. Laskey; Kenneth J. Laskey; Thomas Lukasiewicz; Matthias Nickles; and Michael Pool (eds.), Uncertainty Reasoning for the Semantic Web I, pp. 88–107. Springer-Verlag.

CRISTIANINI, NELLO and SHAWE-TAYLOR, JOHN (2002). An introduction to support vector machines and other kernel-based learning methods. Cambridge Univiversity Press.

CUENCA GRAU, BERNARDO; HORROCKS, IAN; KAZAKOV, YEVGENY; and SATTLER, ULRIKE (2007). A Logical Framework for Modularity of Ontologies. In: Manuela M. Veloso (ed.), 20th International Joint Conference on Artificial Intelligence, pp. 298–303.

CUENCA GRAU, BERNARDO; HORROCKS, IAN; KAZAKOV, YEVGENY; and SATTLER, ULRIKE (2009a). Extracting Modules from Ontologies: A Logic-Based Approach. In: Heiner Stuckenschmidt; Christine Parent; and Stefano Spaccapietra (eds.), Modular Ontologies – Concepts, Theories and Techniques for Knowledge Modularization, pp. 159–186. Springer.

CUENCA GRAU, BERNARDO; PARSIA, BIJAN; and SIRIN, EVREN (2006). Combining OWL ontologies using $\mathcal{E}$-Connections. Journal Of Web Semantics, 4(1), 40–59.

CUENCA GRAU, BERNARDO; PARSIA, BIJAN; and SIRIN, EVREN (2009b). Ontology Integration Using $\mathcal{E}$-connections. In: Heiner Stuckenschmidt; Christine Parent; and Stefano Spaccapietra (eds.), Modular Ontologies – Concepts, Theories and Techniques for Knowledge Modularization, pp. 293–320. Springer.

CUENCA GRAU, BERNARDO; PARSIA, BIJAN; SIRIN, EVREN; and KALYANPUR, ADITYA (2005). Automatic Partitioning of OWL Ontologies Using $\mathcal{E}$-connections. In: Ian Horrocks; Ulrike Sattler; and Frank Wolter (eds.), 18th International Workshop on Description Logics (DL'05). CEUR Workshop Proceedings.

DAVIDSON, DONALD H. (1967). The logical form of action sentences. In: Nicholas Rescher (ed.), The Logic of Decision and Action, pp. 81–95. University of Pittsburgh Press.

DING, ZHONGLI; PENG, YUN; and PAN, RONG (2006). BayesOWL: Uncertainty Modeling in Semantic Web Ontologies. In: Zongmin Ma (ed.), Soft Computing in Ontologies and Semantic Web, vol. 204 of *Studies in Fuzziness and Soft Computing*, pp. 3–29. Springer.

DUBOIS, DIDIER and PRADE, HENRI (1988). Possibility Theory: An Approach to Computerized Processing of Uncertainty (traduction revue et augmente de "Thorie des Possibilits"). Plenum Press.

DUBOIS, DIDIER and PRADE, HENRI (2001). Possibility Theory, Probability Theory and Multiple-Valued Logics: A Clarification. Annals of Mathematics and Artificial Intelligence, 32(1-4), 35–66.

EHRIG, MARC; HAASE, PETER; HEFKE, MARK; and STOJANOVIC, NENAD (2005). Similarity for Ontologies – A Comprehensive Framework. In: 13th European Conference on Information Systems, Information Systems in a Rapidly Changing Economy (ECIS'05).

EL-GERESY, BAHER A. (1997). The Space Algebra: Spatial Reasoning without Composition Tables. IEEE International Conference on Tools with Artificial Intelligence (ICTAI'97), pp. 67–74.

ENSAN, FAEZEH (2008). Formalizing Ontology Modularization through the Notion of Interfaces. In: Aldo Gangemi and Jérôme Euzenat (eds.), Knowledge Engineering and Knowledge Management, pp. 74–82. Springer.

ESSAID, AMIRA and YAGHLANE, BOUTHEINA BEN (2009). BeliefOWL: An Evidential Representation in OWL Ontology. In: Fernando Bobillo; Paulo C. G. da Costa; Claudia d'Amato; Nicola Fanizzi; Kathryn B. Laskey; Kenneth J. Laskey; Thomas Lukasiewicz; Trevor Martin; Matthias Nickles; Michael Pool; and Pavel Smrž (eds.), Fifth Workshop on Uncertainty Reasoning for the Semantic Web (URSW'09), pp. 77–80. CEUR Workshop Proceedings.

EUZENAT, JÉRÔME and SHVAIKO, PAVEL (2007). Ontology Matching. Springer.

EUZENAT, JÉRÔME and VALTCHEV, PETKO (2004). Similarity-based Ontology Alignment in OWL-Lite. In: Ramon López de Mántaras and Lorenza Saitta (eds.), 16th European Conference on Artificial Intelligence (ECAI'04), pp. 333–337. IOS Press.

FAUCONNIER, GILLES and TURNER, MARK (2003). The Way We Think: Conceptual Blending and the Mind's Hidden Complexities. Basic Books.

FERNÁNDEZ-LÓPEZ, MARIANO; GÓMEZ-PÉREZ, ASUNCIÓN; and JURISTO, NATALIA (1997). METHONTOLOGY: from Ontological Art towards Ontological Engineering. In: AAAI'97 Spring Symposium, pp. 33–40.

FONSECA, FREDERICO and RODRIGUEZ, ANDREA (2007). From Geo-Pragmatics to Derivation Ontologies: New Directions for the GeoSpatial Semantic Web. Transactions in GIS, 11(3), 313–316.

FORSYTH, DAVID A. and PONCE, JEAN (2003). Computer Vision: A Modern Approach. Prentice Hall.

FRANK, ANDREW U. (2001). Tiers of ontology and consistency constraints in geographical information systems. International Journal of Geographical Information Science, 15(7), 667–678.

FREKSA, CHRISTIAN (1991). Qualitative Spatial Reasoning. In: David M. Mark and Andrew U. Frank (eds.), Cognitive and Linguistic Aspects of Geographic Space, NATO ASI Series, pp. 361–372. Kluwer.

FREKSA, CHRISTIAN (1992). Using Orientation Information for Qualitative Spatial Reasoning. In: Andrew U. Frank; Irene Campari; and Ubaldo Formentini (eds.), Theories and methods of spatio-temporal reasoning in geographic space, pp. 162–178. Springer.

FROESE, THOMAS; FISCHER, MARTIN; GROBLER, FRANCOIS; RITZENTHALER, JOHN; YU, KEVIN; SUTHERLAND, STUART; STAUB, SHERYL; AKINCI, BURCU; AKBAS, RAGIP; KOO, BONSANG; BARRON, ALEX; and KUNZ, JOHN (1999). Industry Foundation Classes for Project Management - A Trial Implementation. ITCon, 4, 17–36. www.ifcwiki.org/, visited on July 05, 2010.

GALTON, ANTHONY P. (1997). Space, Time, and Movement. In: Oliviero Stock (ed.), Spatial and Temporal Reasoning, pp. 321–352. Kluwer Academic Publishers.

GAMMA, ERICH; HELM, RICHARD; JOHNSON, RALPH; and VLISSIDES, JOHN (1995). Design Patterns: Elements of Reusable Object-Oriented Software. Addison-Wesley.

GANGEMI, ALDO (2010). What's in a schema? A formal metamodel for ECG and FrameNet. In: Chu-Ren Huang; Nicoletta Calzolari; Aldo Gangemi; Alessandro Lenci; Alessandro Oltramari; and Laurent Prévot (eds.), Ontology and the Lexicon – A Natural Language Processing Perspective, Studies in Natural Language Processing, chap. 9, pp. 144–182. Cambridge University Press.

GANGEMI, ALDO; CATENACCI, CAROLA; CIARAMITA, MASSIMILIANO; and LEHMANN, JOS (2005). A theoretical framework for ontology evaluation and validation. In: Proceedings of the 2nd Italian Semantic Web Workshop (SWAP'05). CEUR Workshop Proceedings.

GANGEMI, ALDO and GUARINO, NICOLA (2004). Impact of foundational ontologies on standardization activities. WonderWeb Deliverable D19, ISTC-CNR.

GANGEMI, ALDO and MIKA, PETER (2003). Understanding the Semantic Web through Descriptions and Situations. In: International Conference on Ontologies, Databases and Applications of Semantics (ODBASE'03).

GANTNER, ZENO; WESTPHAL, MATTHIAS; and WÖLFL, STEFAN (2008). GQR - A Fast Reasoner for Binary Qualitative Constraint Calculi. In: AAAI'08 Workshop on Spatial and Temporal Reasoning.

GÄRDENFORS, PETER (2000). Conceptual spaces – the geometry of thought. A Bradford book. MIT Press.

GASSE, FRANCIS; SATTLER, ULRIKE; and HAARSLEV, VOLKER (2008). Rewriting Rules into $\mathcal{SROIQ}$ Axioms. In: Franz Baader; Carsten Lutz; and Boris Motik (eds.), 21st International Workshop on Description Logics (DL'08). CEUR Workshop Proceedings.

GERBER, RALF; NAGEL, HANS-HELLMUT; and SCHREIBER, HEIKO (2002). Deriving Textual Descriptions of Road Traffic Queues from Video Sequences. In: Frank van Harmelen (ed.), 15th Eureopean Conference on Artificial Intelligence (ECAI'02), pp. 736–740. IOS Press.

GÓMEZ-PÉREZ, ASUNCIÓN; FERNÁNDEZ-LÓPEZ, MARIANO; and CORCHO, OSCAR (eds.) (2004). Ontological Engineering – with examples from the areas of Knowledge Management, e-Commerce and the Semantic Web. Springer.

GORDON, JEAN and SHORTLIFFE, EDWARD H. (1985). A method for managing evidential reasoning in a hierarchical hypothesis space. Artificial Intelligence, 26(3), 323–357.

GRENON, PIERRE (2008). A Primer on Knowledge Management and Ontological Engineering. vol. 9 of *Metaphysical Research*, chap. 3, pp. 57–82. Ontos Verlag.

GRENON, PIERRE and SMITH, BARRY (2004). SNAP and SPAN: Towards Dynamic Spatial Ontology. Spatial Cognition and Computation, 4(1), 69–103.

GRUBER, THOMAS R. (1993). Toward Principles for the Design of Ontologies Used for Knowledge Sharing. International Journal Human-Computer Studies, 43(5–6), 907–928.

GRUBER, THOMAS R. (2009). Ontology. In: Ling Liu and M. Tamer Özsu (eds.), Encyclopedia of Database Systems, pp. 1963–1965. Springer.

GRUBINGER, MICHAEL; CLOUGH, PAUL; MÜLLER, HENNING; and DESELAERS, THOMAS (2006). The IAPR Benchmark: A New Evaluation Resource for Visual Information Systems. In: 5th International Conference on Language Resources and Evaluation (LREC'10). ELRA.

GRÜTTER, ROLF; SCHARRENBACH, THOMAS; and BAUER-MESSMER, BETTINA (2008). Improving an RCC-Derived Geospatial Approximation by OWL Axioms. In: Amit P. Sheth;

Steffen Staab; Mike Dean; Massimo Paolucci; Diana Maynard; Tim W. Finin; and Krishnaprasad Thirunarayan (eds.), 7th International Semantic Web Conference (ISWC'08), pp. 293–306. Springer.

GRÜTTER, ROLF; SCHARRENBACH, THOMAS; and WALDVOGEL, BETTINA (2010). Vague Spatio-Thematic Query Processing: A Qualitative Approach to Spatial Closeness. Transactions in GIS, 14(2), 97–109.

GUARINO, NICOLA (1998). Formal Ontology and Information Systems. In: Nicola Guarino (ed.), Formal Ontology in Information Systems (FOIS'98), pp. 3–15. IOS Press.

GUARINO, NICOLA; CARRARA, MASSIMILIANO; and GIARETTA, PIERDANIELE (1994). Formalizing ontological commitments. In: 12th National Conference on Artificial Intelligence (AAAI'94), pp. 560–567. American Association for Artificial Intelligence.

GUARINO, NICOLA and WELTY, CHRISTOPHER (2002). Evaluating ontological decisions with OntoClean. Communications of the ACM, 45(2), 61–65.

GUHA, RAMANATHAN V. and LENAT, DOUGLAS B. (1990). Cyc: A Midterm Report. AI Magazine, Fall(3), 32–59.

HAARSLEV, VOLKER; LUTZ, CARSTEN; and MÖLLER, RALF (1998). Foundations of Spatioterminological Reasoning with Description Logics. In: Anthony G. Cohn; Lenhard K. Schubert; and Stuart C. Shapiro (eds.), 6th International Conference on Principles of Knowledge Representation and Reasoning (KR'98), pp. 112–123. Morgan-Kaufmann Publishers.

HAARSLEV, VOLKER and MÖLLER, RALF (2003). Description Logic Systems with Concrete Domains: Applications for the Semantic Web. In: François Bry; Carsten Lutz; Ulrike Sattler; and Mareike Schoop (eds.), 10th International Workshop on Knowledge Representation meets Databases, vol. 79. CEUR Workshop Proceedings.

HAASE, PETER; LEWEN, HOLGER; STUDER, RUDI; TRAN, DUC THANH; ERDMANN, MICHAEL; D'AQUIN, MATHIEU; and MOTTA, ENRICO (2008). The NeOn Ontology Engineering Toolkit. In: WWW 2008 Developers Track.

HALPERN, JOSEPH Y. (2003). Reasoning about Uncertainty. MIT Press.

HAMEED, ADIL; PREECE, ALUN; and SLEEMAN, DEREK (2004). Ontology reconciliation. In: Handbook of ontologies, International handbooks on information systems, chap. 12, pp. 231–250. Springer.

HAMMER, BARBARA and HITZLER, PASCAL (eds.) (2007). Perspectives of Neural-Symbolic Integration, vol. 77 of Studies in Computational Intelligence. Springer.

HARARY, FRANK (1969). Graph Theory. Addison-Wesley.

HARTIGAN, JOHN A. (1983). Bayes theory. Springer Series in Statistics. Springer.

HAYES, PATRICK J. (1980). The logic of frames. In: Dieter W. Metzing (ed.), Frame Conceptions and Text Understanding, pp. 46–61. deGruyter.

HAYES, PATRICK J. (1985). The Second Naive Physics Manifesto. In: Jerry R. Hobbs and Robert C. Moore (eds.), Formal Theories of the Commonsense World, pp. 1–36. Ablex Publishing Corporation.

HELBIG, HERMANN (2006). Knowledge Representation and the Semantics of Natural Language. Springer.

HENDLER, JAMES (2001). Agents and the Semantic Web. IEEE Intelligent Systems, 16(2), 30–37.

HERSKOVITS, ANNETTE (1986). Language and Spatial Cognition: An interdisciplinary study of the propositions in English. Cambridge University Press.

HERVÁS, RAQUEL; COSTA, RUI P.; COSTA, HUGO; GERVÁS, PABLO; and PEREIRA, FRANCISCO C. (2007). Enrichment of Automatically Generated Texts Using Metaphor. In: Alexander F. Gelbukh and Angel Fernando Kuri Morales (eds.), Advances in Artificial Intelligence, 6th Mexican International Conference on Artificial Intelligence (MICAI'07), vol. 4827 of Lecture Notes in Computer Science, pp. 944–954. Springer.

HILBERT, DAVID (1899). The Foundations of Geometry. The Open Court Publishing Co.

HOIS, JOANA (2006). Modellierung und Klassifikation von Raumkonzepten anhand der Dempster-Shafer-Theorie am Beispiel von Universitätsräumen. Diploma Thesis, Universität Bremen.

HOIS, JOANA (2009). A Semantic Framework for Uncertainties in Ontologies. In: Chad Lane and Hans Guesgen (eds.), Proceedings of the 22nd International FLAIRS conference 2009 (FLAIRS'09). AAAI Press.

HOIS, JOANA (2010a). Formalizing Diverse Spatial Information with Modular Ontologies. In: David N. Rapp (ed.), Poster Presentation at the Spatial Cognition VII, pp. 41–44. SFB/TR 8 Report No. 024-07/2010.

HOIS, JOANA (2010b). Inter-Annotator Agreement on a Linguistic Ontology for Spatial Language – A Case Study for GUM-Space. In: 7th International Conference on Language Resources and Evaluation (LREC'10), pp. 3464–3469. ELRA.

HOIS, JOANA (2010c). Modularizing Spatial Ontologies for Assisted Living Systems. In: Yaxin Bi and Mary-Anne Williams (eds.), 4th International Conference on Knowledge Science, Engineering & Management (KSEM'10), pp. 424–435. Springer.

HOIS, JOANA (2011). Modeling the Diversity of Spatial Information by Using Modular Ontologies and their Combinations. In: Oliver Kutz and Thomas Schneider (eds.), Modular Ontologies: Proceedings of the 5th International Workshop on Modular Ontologies (WoMO'11), pp. 71–78. IOS Press.

HOIS, JOANA; BHATT, MEHUL; and KUTZ, OLIVER (2009a). Modular Ontologies for Architectural Design. In: Roberta Ferrario and Alessandro Oltramari (eds.), 4th Workshop on Formal Ontologies Meet Industry (FOMI'09), vol. 198 of *Frontiers in Artificial Intelligence and Applications*. IOS Press.

HOIS, JOANA; DYLLA, FRANK; and BHATT, MEHUL (2009b). Qualitative Spatial and Terminological Reasoning for Ambient Environments - Recent Trends and Future Directions. In: Workshop on Spatial and Temporal Reasoning for Ambient Intelligence Systems (STAMI'09).

HOIS, JOANA and KUTZ, OLIVER (2008a). Counterparts in Language and Space - Similarity and *S*-Connection. In: Formal Ontology in Information Systems (FOIS'08), pp. 266–279. IOS Press.

HOIS, JOANA and KUTZ, OLIVER (2008b). Natural Language meets Spatial Calculi. In: Christian Freksa; Nora S. Newcombe; Peter Gärdenfors; and Stefan Wölfl (eds.), Spatial Cognition VI – Learning, Reasoning, and Talking about Space, pp. 266–282. Springer.

HOIS, JOANA; KUTZ, OLIVER; and BATEMAN, JOHN A. (2008). Similarity-Connections between Natural Language and Spatial Situations. In: Spatial Language in Context: Computational and Theoretical Approaches to Situation Specific Meaning, Workshop at Spatial Cognition VI.

HOIS, JOANA; KUTZ, OLIVER; MOSSAKOWSKI, TILL; and BATEMAN, JOHN (2010). Towards Ontological Blending. In: Darina Dicheva and Danail Dochev (eds.), 14th International Conference on Artificial Intelligence: Methodology, Systems, Applications (AIMSA'10), pp. 263–264. Poster Presentation.

HOIS, JOANA; SCHILL, KERSTIN; and BATEMAN, JOHN A. (2006). Integrating Uncertain Knowledge in a Domain Ontology for Room Concept Classifications. In: Max Bramer; Frans Coenen; and Andrew Tuson (eds.), The Twenty-sixth SGAI International Conference on Innovative Techniques and Applications of Artificial Intelligence, Research and Development in Intelligent Systems XXIII, pp. 245–258. Springer.

HOIS, JOANA; TENBRINK, THORA; ROSS, ROBERT; and BATEMAN, JOHN (2009c). GUM-Space – The Generalized Upper Model spatial extension: A linguistically-motivated ontology for the semantics of spatial language. Tech. rep., Collaborative Research Center for Spatial Cognition, University of Bremen, Germany.

HOIS, JOANA; WÜNSTEL, MICHAEL; BATEMAN, JOHN A.; and RÖFER, THOMAS (2007). Dialog-Based 3D-Image Recognition Using a Domain Ontology. In: Thomas Barkowsky;

Markus Knauff; Gérard Ligozat; and Daniel R. Montello (eds.), Spatial Cognition V: Reasoning, Action, Interaction: International Conference Spatial Cognition 2006. Springer.

HOLI, MARKUS and HYVÖNEN, EERO (2005). Modeling Degrees of Conceptual Overlap in Semantic Web Ontologies. In: Paulo C. G. da Costa; Kathryn B. Laskey; Kenneth J. Laskey; and Michael Pool (eds.), Workshop on Uncertainty Reasoning for the Semantic Web (URSW'05), pp. 98–99. CEUR Workshop Proceedings.

HORRIDGE, MATTHEW and BECHHOFER, SEAN (2009). The OWL API: A Java API for Working with OWL 2 Ontologies. In: Rinke Hoekstra and Peter F. Patel-Schneider (eds.), 6th International Workshop on OWL: Experiences and Directions (OWLED'09).

HORRIDGE, MATTHEW and PATEL-SCHNEIDER, PETER F. (2008). Manchester OWL Syntax for OWL 1.1. In: OWL: Experiences and Directions.

HORROCKS, IAN; KUTZ, OLIVER; and SATTLER, ULRIKE (2006). The Even More Irresistible $\mathcal{SROIQ}$. In: Knowledge Representation and Reasoning. AAAI Press.

HORROCKS, IAN; PATEL-SCHNEIDER, PETER F.; BOLEY, HAROLD; TABET, SAID; GROSOF, BENJAMIN; and DEAN, MIKE (2004). SWRL: A Semantic Web Rule Language – Combining OWL and RuleML. Tech. rep., W3C. http://www.w3.org/Submission/SWRL/, visited on July 04, 2011.

HUDELOT, CÉLINE; ATIF, JAMAL; and BLOCH, ISABELLE (2008). Fuzzy spatial relation ontology for image interpretation. Fuzzy Sets and Systems, 159(15), 1929–1951.

IPFELKOFER, FRANK; LORENZ, BERNHARD; and OHLBACH, HANS JÜRGEN (2006). Ontology Driven Visualisation of Maps with SVG – An Example for Semantic Programming. In: 10th International Conference on Information Visualisation (IV'06), pp. 424–429. IEEE Computer Society.

JANOWICZ, KRZYSZTOF (2006). Sim-DL: Towards a Semantic Similarity Measurement Theory for the Description Logic ALCNR in Geographic Information Retrieval. In: On the Move to Meaningful Internet Systems 2006: OTM 2006 Workshops, pp. 1681–1692. Springer.

JANSEN, LUDGER (2008a). Categories: The Top-Level Ontology. vol. 9 of Metaphysical Research, chap. 8, pp. 173–196. Ontos Verlag.

JANSEN, LUDGER (2008b). Classifications. vol. 9 of Metaphysical Research, chap. 7, pp. 159–172. Ontos Verlag.

JOHNSON, MARK (1987). The Body in the Mind: The bodily basis of meaning, imagination, and reason. University of Chicago Press.

JOHNSON-LAIRD, PHILIP N. (1983). Mental models: towards a cognitive science of language, inference, and consciousness. Cambridge University Press.

JOHNSTON, BENJAMIN; YANG, FANGKAI; MENDOZA, ROGAN; CHEN, XIAOPING; and WILLIAMS, MARY-ANNE (2008). Ontology Based Object Categorization for Robots. In: Takahira Yamaguchi (ed.), 7th International Conference on Practical Aspects of Knowledge Management (PAKM'08), vol. 5345 of *Lecture Notes in Computer Science*, pp. 219–231. Springer.

JUPP, JULIE R. and GERO, JOHN S. (2006). Towards computational analysis of style in architectural design. Journal of the American Society for Information Science, 57(11), 1537–1550.

KALAY, YEHUDA E. and MITCHELL, WILLIAM J. (2004). Architecture's New Media: Principles, Theories, and Methods of Computer-Aided Design. The MIT Press.

KATZ, BORIS; LIN, JIMMY; STAUFFER, CHRIS; and GRIMSON, ERIC (2004). Answering Questions About Moving Objects in Videos. In: Mark T. Maybury (ed.), New Directions in Question Answering, pp. 113–128. MIT Press.

KEHAGIAS, DIONYSIOS; KONTOTASIOU, DIONYSIA; MOURATIDIS, GEORGIOS; NIKOLAOU, THEOFILOS; PAPADIMITRIOU, IOANNIS; KALOGIROU, KOSTAS; BATEMAN, JOHN; GARCIA, ALEXANDER; and NORMANN, IMMANUEL (2009). Ontologies, typologies, models and management tools. OASIS Project Deliverable D1.1.1, CERTH and University of Bremen.

KEHAGIAS, DIONYSIOS D.; PAPADIMITRIOU, IOANNIS; HOIS, JOANA; TZOVARAS, DIMITRIOS; and BATEMAN, JOHN A. (2008). A Methodological Approach for Ontology Evaluation and Refinement. In: ASK-IT International Conference 2008.

KELLEHER, JOHN D. and COSTELLO, FINTAN J. (2009). Applying Computational Models of Spatial Prepositions to Visually Situated Dialog. Computational Linguistics, 35(2), 271–306.

KELLEHER, JOHN D. and KRUIJFF, GEERT-JAN M. (2006). Incremental Generation of Spatial Referring Expressions in Situated Dialogue. In: Joint conference of the International Committee on Computational Linguistics and the Association for Computational Linguistics (Coling/ACL'06).

KJELLSTRÖM, HEDVIG; ROMERO, JAVIER; MARTÍNEZ, DAVID; and KRAGIĆ, DANICA (2008). Simultaneous Visual Recognition of Manipulation Actions and Manipulated Objects. In: David Forsyth; Philip Torr; and Andrew Zisserman (eds.), 10th European Conference on Computer Vision (ECCV'08), pp. 336–349. Springer.

KLINOV, PAVEL (2008). Pronto: A Non-monotonic Probabilistic Description Logic Reasoner. In: Sean Bechhofer; Manfred Hauswirth; Jörg Hoffmann; and Manolis Koubarakis (eds.), The Semantic Web: Research and Applications, pp. 822–826. Springer.

KLIPPEL, ALEXANDER; FREKSA, CHRISTIAN; and WINTER, STEPHAN (2006). You-Are-Here Maps in Emergencies – The Danger of Getting Lost. Journal of Spatial Science, 51(1).

KNAUFF, MARKUS (1997). Räumliches Wissen und Gedächtnis. Deutscher Universitätsverlag.

KÖHLER, JACOB; MUNN, KATHERINE; RÜEGG, ALEXANDER; SKUSA, ANDRE; and SMITH, BARRY (2006). Quality control for terms and definitions in ontologies and taxonomies. BMC Bioinformatics, 7(212).

KONEV, BORIS; LUTZ, CARSTEN; WALTHER, DIRK; and WOLTER, FRANK (2009). Formal Properties of Modularisation. In: Heiner Stuckenschmidt; Christine Parent; and Stefano Spaccapietra (eds.), Modular Ontologies – Concepts, Theories and Techniques for Knowledge Modularization, pp. 25–66. Springer.

KORDJAMSHIDI, PARISA; HOIS, JOANA; VAN OTTERLO, MARTIJN; and MOENS, MARIE-FRANCINE (2011a). Machine Learning for Interpretation of Spatial Natural Language in terms of QSR. In: Nicholas Giudice; Mike Worboys; Max Egenhofer; and Reinhardt Moratz (eds.), 10th International Conference on Spatial Information Theory (COSIT'11). Poster Presentation.

KORDJAMSHIDI, PARISA; HOIS, JOANA; VAN OTTERLO, MARTIJN; and MOENS, MARIE-FRANCINE (2013). Learning to Interpret Spatial Natural Language in terms of Qualitative Spatial Relations. In: Thora Tenbrink; Jan Wiener; and Christophe Claramunt (eds.), Representing space in cognition: Interrelations of behavior, language, and formal models. Oxford University Press. In press.

KORDJAMSHIDI, PARISA; VAN OTTERLO, MARTIJN; and MOENS, MARIE-FRANCINE (2011b). Spatial role labeling: Towards extraction of spatial relations from natural language. ACM Transactions on Speech and Language Processing, 8(3), 4:1–4:36.

KOSSLYN, STEPHEN M. and POMERANTZ, JAMES R. (1977). Imagery, propositions and the form of internal representations. Cognitive Psychology, 9, 52–76.

KOVACS, KATALIN; DOLBEAR, CATHERINE; and GOODWIN, JOHN (2007). Spatial Concepts and OWL issues in a topographic ontology framework. In: Geographic Information Science Conference (GISRUK'07).

KRACHT, MARCUS (2008). Language and Space. Book manuscript.

KRIEG-BRÜCKNER, BERND; GERSDORF, BERND; DÖHLE, MATHIAS; and SCHILL, KERSTIN (2009). Technology for Seniors to Be in the Bremen Ambient Assisted Living Lab. In: 2. Deutscher AAL-Kongress. VDE-Verlag.

KRIEG-BRÜCKNER, BERND and SHI, HUI (2006). Orientation Calculi and Route Graphs: Towards Semantic Representations for Route Descriptions. In: Martin Raubal; Harvey J. Miller; Andrew U. Frank; and Michael F. Goodchild (eds.), International Conference on Geographic Information Science (GIScience'06), pp. 234–250. Springer.

KRIEG-BRÜCKNER, BERND and SHI, HUI (2009). Spatio-Temporal Situated Interaction in Ambient Assisted Living. In: Helge Ritter; Gerhard Sagerer; and Jochen Steil (eds.), Third International Workshop on Human Centered Robotic Systems (HCRS'09), vol. 6 of *Cognitive Systems Monographs*. Springer.

KRUIJFF, GEERT-JAN M.; ZENDER, HENDRIK; JENSFELT, PATRIC; and CHRISTENSEN, HENRIK I. (2007). Situated dialogue and spatial organization: What, where. . . and why? International Journal of Advanced Robotic Systems, 4(1), 125–138.

KUHN, WERNER (2002). Modeling the Semantics of Geographic Categories through Conceptual Integration. In: Max J. Egenhofer and David M. Mark (eds.), 2nd International Conference on Geographic Information Science (GIScience'02), pp. 108–118. Springer.

KUHN, WERNER and FRANK, ANDREW U. (1991). A formalization of metaphors and image-schemas in user interfaces. In: David M. Mark and Andrew U. Frank (eds.), Cognitive and Linguistic Aspects of Geographic Space, NATO ASI Series, pp. 419–434. Kluwer.

KUIPERS, BENJAMIN (1978). Modeling Spatial Knowledge. Cognitive Science, 2(2), 129–153.

KURATA, YOHEI and EGENHOFER, MAX (2007). The 9^+-Intersection for Topological Relations between a Directed Line Segment and a Region. In: Björn Gottfried (ed.), 1st Workshop on Behaviour Monitoring and Interpretation, pp. 62–76.

KURATA, YOHEI and SHI, HUI (2008). Interpreting Motion Expressions in Route Instructions Using Two Projection-Based Spatial Models. In: KI 2008: Advances in Artificial Intelligence, pp. 258–266.

KUTZ, OLIVER (2004). $\mathcal{E}$-Connections and Logics of Distance. Ph.D. thesis, University of Liverpool.

KUTZ, OLIVER; HOIS, JOANA; BAO, JIE; and CUENCA GRAU, BERNARDO (eds.) (2010a). Modular Ontologies – Proceedings of the Fourth International Workshop (WoMO'10). Frontiers in Artificial Intelligence and Applications. IOS Press.

KUTZ, OLIVER; LÜCKE, DOMINIK; and MOSSAKOWSKI, TILL (2008). Heterogeneously Structured Ontologies—Integration, Connection, and Refinement. In: Thomas Meyer and Mehmet A. Orgun (eds.), Knowledge Representation Ontology Workshop, pp. 41–50. ACS.

KUTZ, OLIVER; LUTZ, CARSTEN; WOLTER, FRANK; and ZAKHARYASCHEV, MICHAEL (2004). $\mathcal{E}$-Connections of Abstract Description Systems. Artificial Intelligence, 156(1), 1–73.

KUTZ, OLIVER; MOSSAKOWSKI, TILL; and LÜCKE, DOMINIK (2010b). Carnap, Goguen, and the Hyperontologies: Logical Pluralism and Heterogeneous Structuring in Ontology Design. Logica Universalis, 4, 255–333.

LAAMARI, NAJOUA and BEN YAGHLANE, BOUTHEINA (2007). Uncertainty in Semantic Ontology Mapping: An Evidential Approach. In: Khaled Mellouli (ed.), Symbolic and Quantitative Approaches to Reasoning with Uncertainty (ECSQARU'07), pp. 418–429. Springer.

LAKOFF, GEORGE (1990). The Invariance Hypothesis: is abstract reason based on image-schemas? Cognitive Linguistics, 1(1), 39–74.

LANG, EWALD (1991). The LILOG Ontology from a Linguistic Point of View. In: Otthein Herzog and Claus-Rainer Rollinger (eds.), Text Understanding in LILOG, pp. 464–481. Springer.

LAWSON, BRYAN (2001). The Language of Space. Architectural Press.

LEFEBVRE, HENRI (1991). The production of space. Blackwell.

LEHAR, STEVEN (2004). Gestalt Isomorphism and the Primacy of Subjective Conscious Experience: A Gestalt Bubble Model. Behavioral & Brain Sciences, 26(4), 375–444.

LENAT, DOUGLAS B. and GUHA, RAMANATHAN V. (1990). Building Large Knowledge-Based Systems: Representation and Inference in the Cyc Project. Addison-Wesley.

LEVINSON, STEPHEN C. (2003). Space in Language and Cognition: Explorations in Cognitive Diversity. Cambridge University Press.

LIAO, JING; BI, YAXIN; and NUGENT, CHRIS (2010). Engineering Knowledge for Assistive Living. In: Yaxin Bi and Mary-Anne Williams (eds.), 4th International Conference on Knowledge Science, Engineering and Management (KSEM'10), pp. 186–197. Springer.

LIEBICH, THOMAS; ADACHI, YOSHINOBU; FORESTER, JAMES; HYVARINEN, JUHA; RICHTER, STEFAN; CHIPMAN, TIM; WEISE, MATTHIAS; and WIX, JEFFREY (2010). Industry Foundation Classes – IFC2x Edition 4 Release Candidate 2. Model Support Group (MSG).

LOUX, MICHAEL J. (2002). Metaphysics – a contemporary introduction. Routledge Contemporary Introductions to Philosophy. Routledge, 2nd edn.

LOWE, E. JONATHAN (2010). Action Theory and Ontology. In: Timothy O'Connor and Constantine Sandis (eds.), A Companion to the Philosophy of Action, pp. 3–9. Blackwell Publishing Ltd.

LUKASIEWICZ, THOMAS and STRACCIA, UMBERTO (2008). Managing uncertainty and vagueness in description logics for the Semantic Web. Journal of Web Semantics, 6, 291–308.

LYNCH, KEVIN A. (1960). The image of the city. MIT Press.

MACMAHON, MATT; STANKIEWICZ, BRIAN; and KUIPERS, BENJAMIN (2006). Walk the Talk: Connecting Language, Knowledge, and Action in Route Instructions. In: 21st National Conference on Artificial Intelligence (AAAI'06). AAAI Press.

MAEDCHE, ALEXANDER and STAAB, STEFFEN (2002). Measuring Similarity between Ontologies. In: 13th International Conference on Knowledge Engineering and Knowledge Management. Ontologies and the Semantic Web (EKAW'02), pp. 251–263. Springer.

MAILLOT, NICOLAS ERIC and THONNAT, MONIQUE (2008). Ontology Based Complex Object Recognition. Image and Vision Computing, 26(1), 102–113.

MANI, INDERJEET; HITZEMAN, JANET; RICHER, JUSTIN; HARRIS, DAVE; QUIMBY, ROB; and WELLNER, BEN (2008). SpatialML: Annotation Scheme, Corpora, and Tools. In: 6th Int. Language Resources and Evaluation, LREC'08.

MANN, WILLIAM C. (ed.) (2005). Smart Technology for Aging, Disability, and Independence: The State of the Science. John Wiley & Sons.

MARK, DAVID M. and SMITH, BARRY (2004). A Science of Topography: Bridging the Qualitative-Quantitative Divide. In: Michael P. Bishop and John F. Shroder (eds.), Geographic Information Science and Mountain Geomorphology, pp. 75–100. Springer.

MASOLO, CLAUDIO; BORGO, STEFANO; GANGEMI, ALDO; GUARINO, NICOLA; and OLTRAMARI, ALESSANDRO (2003). Ontologies library. WonderWeb Deliverable D18, ISTC-CNR.

MCNEILL, FIONA (2006). Dynamic Ontology Refinement. Ph.D. thesis, University of Edinburgh.

MINSKY, MARVIN (1974). A Framework for Representing Knowledge. Tech. rep., MIT-AI Laboratory Memo 306.

MITRA, PRASENJIT; NOY, NATALYA F.; and JAISWAL, ANUJ R. (2004). OMEN: A Probabilistic Ontology Mapping Tool. In: Workshop on Meaning Coordination and Negotiation, pp. 537–547.

MITRA, PRASENJIT; NOY, NATASHA; and JAISWAL, ANUJ R. (2005). Ontology Mapping Discovery with Uncertainty. In: Yolanda Gil; Enrico Motta; V. Richard Benjamins; and Mark A. Musen (eds.), International Semantic Web Conference (ISWC'05), pp. 537–547. Springer.

MONTELLO, DANIEL R. (1995). How significant are cultural differences in spatial cognition? In: Andrew U. Frank and Werner Kuhn (eds.), Spatial information theory: A theoretical basis for GIS (COSIT'95), pp. 485–500. Springer.

MONTELLO, DANIEL R. (1998). Kartenverstehen: Die Sicht der Kognitionspsychologie [Understanding maps: The view from cognitive psychology]. Zeitschrift für Semiotik, 20(1-2), 91–103.

MORATZ, REINHARD (2006). Representing Relative Direction as a Binary Relation of Oriented Points. In: Gerhard Brewka; Silvia Coradeschi; Anna Perini; and Paolo Traverso (eds.), 17th European Conference on Artificial Intelligence (ECAI'06), pp. 407–411. IOS Press.

MORATZ, REINHARD and TENBRINK, THORA (2006). Spatial reference in linguistic human-robot interaction: Iterative, empirically supported development of a model of projective relations. Spatial Cognition and Computation, 6(1), 63–107.

MOSSAKOWSKI, TILL; HAXTHAUSEN, ANNE; SANNELLA, DON; and TARLECKI, ANDRZEJ (2008). CASL, the Common Algebraic Specification Language. In: Dines Bjørner and Martin C. Henson (eds.), Logics of formal specification languages, Monographs in Theoretical Computer Science, chap. 3, pp. 241–298. Springer-Verlag Heidelberg.

MOSSAKOWSKI, TILL; LANGE, CHRISTOPH; and KUTZ, OLIVER (2012). Three Semantics for the Core of the Distributed Ontology Language. In: Maureen Donnelly and Giancarlo Guizzardi (eds.), Formal Ontology in Information Systems, Proceedings of the 7th International Conference (FOIS'12), pp. 337–352. IOS Press.

MOSSAKOWSKI, TILL; MAEDER, CHRISTIAN; and LÜTTICH, KLAUS (2007). The Heterogeneous Tool Set. In: Orna Grumberg and Michael Huth (eds.), 13th International Conference on Tools and Algorithms for the Construction and Analysis of Systems (TACAS'07), pp. 519–522. Springer.

MOTIK, BORIS; PATEL-SCHNEIDER, PETER F.; and CUENCA GRAU, BERNARDO (2008). OWL 2 Web Ontology Language: Direct Semantics. Tech. rep., W3C. http://www.w3.org/TR/owl2-semantics/, visited on July 05, 2010.

MOTIK, BORIS; SHEARER, ROB; and HORROCKS, IAN (2009). Hypertableau Reasoning for Description Logics. Journal of Artificial Intelligence Research, 36, 165–228.

MUNN, KATHERINE and SMITH, BARRY (eds.) (2008). Applied Ontology, vol. 9 of *Metaphysical Research*. Ontos Verlag.

NAGY, MIKLOS; VARGAS-VERA, MARIA; and MOTTA, ENRICO (2007). DSSim - managing uncertainty on the Semantic Web. In: Pavel Shvaiko; Jérôme Euzenat; Fausto Giunchiglia; and Bin He (eds.), 2nd International Workshop on Ontology Matching (OM'07).

NEUHAUS, FABIAN; GRENON, PIERRE; and SMITH, BARRY (2004). A Formal Theory of Substances, Qualities, and Universals. In: Achille Varzi and Laure Vieu (eds.), Formal Ontology in Information Systems (FOIS'04), pp. 49–59. IOS Press.

NEUHAUS, FABIAN and HAYES, PAT (2012). Common Logic and the Horatio Problem. Applied Ontology, 7(2), 211–231.

NEUMANN, BERND and MÖLLER, RALF (2004). On Scene Interpretation with Description Logics. Tech. rep., Universität Hamburg. FBI-B-257/04.

NEWCOMBE, NORA S. and RATLIFF, KRISTIN R. (2007). Explaining the Development of Spatial Reorientation – Modularity-Plus-Language versus the Emergence of Adaptive Combination. In: Jodie M. Plumert and John P. Spencer (eds.), The Emerging Spatial Mind, pp. 53–76. Oxford University Pres.

NEWELL, ALLEN and SIMON, HERBERT A. (1976). Computer science as empirical inquiry: symbols and search. Communications of the ACM, 19(3), 113–126.

NICOLA, ANTONIO DE; MISSIKOFF, MICHELE; and NAVIGLI, ROBERTO (2009). A software engineering approach to ontology building. Information Systems, 34(2), 258–275.

NILES, IAN and PEASE, ADAM (2001). Towards a Standard Upper Ontology. In: Christopher Welty and Barry Smith (eds.), Formal Ontology in Information Systems (FOIS'01), pp. 2–9. ACM Press.

NORMANN, IMMANUEL (2009). Automated Theory Interpretation. Ph.D. thesis, Department of Computer Science, Jacobs University, Bremen.

NORMANN, IMMANUEL; DYLLA, FRANK; HOIS, JOANA; KUTZ, OLIVER; BHATT, MEHUL; SCHMITT, MARIO; PUTZ, WOLFGANG; and WEBER, SEBASTIAN (2009). Ontological and Spatial Modeling of Situational Contexts for Ambient Assisted Living. In: OASIS 1st International Conference.

NOY, NATALYA F. and MUSEN, MARK A. (2002). Evaluating Ontology-Mapping Tools: Requirements and Experience. In: Jürgen Angele and York Sure (eds.), Workshop on Evaluation of Ontology Tools (EON'02), pp. 1–14.

OBRST, LEO; CEUSTERS, WERNER; MANI, INDERJEET; RAY, STEVE; and SMITH, BARRY (2007). The Evaluation of Ontologies – Toward Improved Semantic Interoperability. In: Christopher J. O. Baker and Kei-Hoi Cheung (eds.), Semantic Web – Revolutionizing Knowledge Discovery in the Life Sciences, chap. 7, pp. 139–158. Springer.

OLIVA, AUDE and TORRALBA, ANTONIO (2001). Modeling the Shape of the Scene: A Holistic Representation of the Spatial Envelope. International Journal of Computer Vision, 42(3), 145–175.

OLIVÉ, ANTONI (2007). Conceptual Modeling of Information Systems. Springer.

OLSON, DAVID R. and BIALYSTOK, ELLEN (1983). Spatial cognition: the structure and development of mental representations of spatial relations. Child Psychology. Lawrence Erlbaum Associates.

OLTRAMARI, ALESSANDRO; GANGEMI, ALDO; HUANG, CHU-REN; CALCOLARI, NICOLETTA; LENCI, ALESSANDRO; and PRÉVOT, LAURENT (2010). Synergizing ontologies and the lexicon: a roadmap. In: Chu-Ren Huang; Nicoletta Calzolari; Aldo Gangemi; Alessandro Lenci; Alessandro Oltramari; and Laurent Prévot (eds.), Ontology and the Lexicon – A Natural Language Processing Perspective, Studies in Natural Language Processing, chap. 5, pp. 72–78. Cambridge University Press.

Ó NUALLÁIN, SEÁN (ed.) (2000). Spatial Cognition – Foundations and Applications, vol. 26 of *Advances in Consciousness Research*. John Benjamins.

PALMA, RAÚL; HARTMANN, JENS; and HAASE, PETER (2009). OMV - Ontology Metadata Vocabulary for the SemanticWeb. Tech. Rep. v2.4.1, OMV Consortium.

PAN, RONG; DING, ZHONGLI; YU, YANG; and PENG, YUN (2005). A Bayesian Network Approach to Ontology Mapping. In: Yolanda Gil; Enrico Motta; V. Richard Benjamins; and Mark A. Musen (eds.), 4th International Semantic Web Conference (ISWC'05), pp. 563–577. Springer.

PARENT, CHRISTINE and SPACCAPIETRA, STEFANO (2009). An Overview of Modularity. In: Heiner Stuckenschmidt; Christine Parent; and Stefano Spaccapietra (eds.), Modular Ontologies – Concepts, Theories and Techniques for Knowledge Modularization, pp. 5–23. Springer.

PEREIRA, FRANCISCO CÁMARA (2007). Creativity and Artificial Intelligence: A Conceptual Blending Approach, vol. 4 of *Applications of Cognitive Linguistics (ACL)*. Mouton de Gruyter.

PIAGET, JEAN and INHELDER, BÄRBEL (1997). Selected works Vol. 4: The child's conception of space. Routledge, Repr. London 1956 edn.

PIKE, WILLIAM and GAHEGAN, MARK (2007). Beyond ontologies: Toward situated representations of scientific knowledge. International Journal of Man-Machine Studies, 65(7), 674–688.

PORZEL, ROBERT and MALAKA, RAINER (2004). A task-based approach for ontology evaluation. In: Workshop on Ontology Learning and Population at ECAI'04.

PROBST, FLORIAN (2006). Ontological Analysis of Observations and Measurements. In: Martin Raubal; Harvey J. Miller; Andrew U. Frank; and Michael F. Goodchild (eds.), Geographic Information Science, 4th International Conference (GIScience'06), pp. 304–320. Springer.

QI, GUILIN; JI, QIU; PAN, JEFF; and DU, JIANFENG (2010). PossDL – A Possibilistic DL Reasoner for Uncertainty Reasoning and Inconsistency Handling. In: Lora Aroyo; Grigoris Antoniou; Eero Hyvnen; Annette ten Teije; Heiner Stuckenschmidt; Liliana Cabral; and Tania Tudorache (eds.), The Semantic Web: Research and Applications, pp. 416–420. Springer.

RACER SYSTEMS (2007). RacerPro User's Guide. Racer Systems GmbH & Co. KG. Version 1.9.2.

RAMOS, CARLOS (2007). Ambient Intelligence - A State of the Art from Artificial Intelligence Perspective. In: José Neves; Manuel Filipe Santos; and José Machado (eds.), 13th Portuguese Conference on Aritficial Intelligence, pp. 285–295. Springer.

RANDELL, DAVID A.; CUI, ZHAN; and COHN, ANTHONY G. (1992). A spatial logic based on regions and connection. In: Third International Conference on Knowledge Representation and Reasoning, pp. 165–176. Morgan Kaufmann, San Mateo.

RECTOR, ALAN L. (2003). Modularisation of domain ontologies implemented in description logics and related formalisms including OWL. In: Second International Conference on Knowledge Capture (K-CAP'03), pp. 121–128. ACM.

REINEKING, THOMAS; SCHULT, NICLAS; and HOIS, JOANA (2009). Evidential Combination of Ontological and Statistical Information for Active Scene Classification. In: International Conference on Knowledge Engineering and Ontology Development (KEOD'09).

REINEKING, THOMAS; SCHULT, NICLAS; and HOIS, JOANA (2011). Combining Statistical and Symbolic Reasoning for Active Scene Categorization. In: Knowledge Discovery, Knowledge Engineering and Knowledge Management, Revised Selected Papers, pp. 262–275. Springer.

REITER, RAYMOND (1980). A logic for default reasoning. Artificial Intelligence, 13, 81–132.

RENZ, JOCHEN (2002). Qualitative Spatial Reasoning with Topological Information. Springer.

RENZ, JOCHEN and NEBEL, BERNHARD (2007). Qualitative Spatial Reasoning Using Constraint Calculi. In: Marco Aiello; Ian Pratt-Hartmann; and Johan van Benthem (eds.), Handbook of Spatial Logics, pp. 161–215. Springer.

RENZ, JOCHEN; RAUH, REINHOLD; and KNAUFF, MARKUS (2000). Towards Cognitive Adequacy of Topological Spatial Relations. In: Christian Freksa; Wilfried Brauer; Christopher Habel; and Karl Friedrich Wender (eds.), Spatial Cognition II, pp. 184–197. Springer.

RICHENS, RICHARD H. (1956). Preprogramming for mechanical translation. Mechanical Translation, 3(1), 20–25.

ROSS, ROBERT (2008). Tiered Models of Spatial Language Interpretation. In: Christian Freksa; Nora Newcombe; Peter Gärdenfors; and Stefan Wölfl (eds.), Spatial Cognition VI. Learning, Reasoning, and Talking about Space, pp. 233–249. Springer.

ROY, DEB and REITER, EHUD (2005). Connecting Language to the World. Artificial Intelligence, 167(1–2), 1–12.

RUBIN, DANIEL L.; NOY, NATALYA F.; and MUSEN, MARK A. (2007). Protégé: A Tool for Managing and Using Terminology in Radiology Applications. Journal of Digital Imaging, 20(1), 34–46.

RUSSELL, BRYAN C.; TORRALBA, ANTONIO; MURPHY, KEVIN P.; and FREEMAN, WILLIAM T. (2008). LabelMe: a database and web-based tool for image annotation. International Journal of Computer Vision, 77(1–3), 157–173.

SALTON, GERARD (1989). Automatic Text Processing: The Transformation, Analysis, and Retrieval of Information by Computer. Addison-Wesley.

SANDEWALL, ERIK (2006). Ontology of Actions.

SCHILL, KERSTIN (1997). Decision support systems with adaptive reasoning strategies. In: Christian Freksa; Matthias Jantzen; and Rüdiger Valk (eds.), Foundations of Computer Sciences: Theory, Cognition, Application, pp. 417–427. Springer.

SCHILL, KERSTIN; UMKEHRER, ELISABETH; BEINLICH, STEPHAN; KRIEGER, GERHARD; and ZETZSCHE, CHRISTOPH (2001). Scene analysis with saccadic eye movements: top-down and bottom-up modelling. Journal of Electronic Imaging, 10(1), 152–160.

SCHILL, KERSTIN; ZETZSCHE, CHRISTOPH; and HOIS, JOANA (2009). A Belief-Based Architecture for Scene Analysis: from Sensorimotor Features to Knowledge and Ontology. Fuzzy Sets and Systems, 160(10), 1507–1516.

SCHULTZ, CARL and BHATT, MEHUL (2010). A Multi-Modal Data Access Framework for Spatial Assistance Systems. In: Second ACM SIGSPATIAL International Workshop on Indoor Spatial Awareness (ISA'10). ACM.

SCOTT, KENDALL (2001). The Unified Process Explained. Addison-Wesley Professional.

SERAFINI, LUCIANO and HOMOLA, MARTIN (2010). Modular Knowledge Representation and Reasoning in the Semantic Web. In: Roberto de Virgilio; Fausto Giunchiglia; and Letizia Tanca (eds.), Semantic Web Information Management, pp. 147–181. Springer.

SERAFINI, LUCIANO and TAMILIN, ANDREI (2005). DRAGO: Distributed Reasoning Architecture for the Semantic Web. In: Asunción Gómez-Pérez and Jérôme Euzenat (eds.), European Semantic Web Conference (ESWC'05), pp. 361–376. Springer.

SHAFER, GLENN (1976). A Mathematical Theory of Evidence. Princeton University Press.

SHEREMET, MIKHAIL; TISHKOVSKY, DMITRY; WOLTER, FRANK; and ZAKHARYASCHEV, MICHAEL (2007). A Logic for Concepts and Similarity. Journal of Logic and Computation, 17(3), 415–452.

SHI, HUI and TENBRINK, THORA (2009). Telling Rolland where to go: HRI dialogues on route navigation. In: Kenny Coventry; Thora Tenbrink; and John Bateman (eds.), Spatial Language and Dialogue, pp. 177–189. Oxford Univ. Press.

SHVAIKO, PAVEL; GUINCHIGLIA, FAUSTO; and YATSKEVICH, MIKALAI (2010). Semantic Matching with S-Match. In: Roberto de Virgilio; Fausto Giunchiglia; and Letizia Tanca (eds.), Semantic Web Information Management, pp. 183–202. Springer.

SIRIN, EVREN; PARSIA, BIJAN; CUENCA GRAU, BERNARDO; KALYANPUR, ADITYA; and KATZ, YARDEN (2007). Pellet: A practical OWL-DL reasoner. Web Semantics, 5(2), 51–53.

SMITH, ARNOLD (2000). Spatial Cognition Without Spatial Concepts. In: Seán Ó Nualláin (ed.), Spatial Cognition – Foundations and Applications, vol. 26 of *Advances in Consciousness Research*, pp. 127–136. John Benjamins.

SMITH, BARRY and VARZI, ACHILLE C. (1999). The Niche. Noûs, 33(2), 214–238.

SMITHSON, MICHAEL (1989). Ignorance and Uncertainty – Emerging Paradigms. Springer.

SOWA, JOHN F. (1992). Semantic networks. In: Stuart C. Shapiro (ed.), Encyclopedia of Artificial Intelligence. Wiley, 2nd edn. Revised version from 1987.

STAHL, CHRISTOPH and SCHWARTZ, TIM (2010). Modeling and simulating assistive environments in 3-D with the YAMAMOTO toolkit. In: International Conference on Indoor Positioning and Indoor Navigation, pp. 1–6. IEEE Xplore.

STOILOS, GIORGOS; STAMOU, GIORGOS; and PAN, JEFF Z. (2010). Fuzzy extensions of OWL: Logical properties and reduction to fuzzy description logics. International Journal of Approximate Reasoning, 51(6), 656–679.

STOILOS, GIORGOS; STAMOU, GIORGOS; TZOUVARAS, VASSILIS; PAN, JEFF Z.; and HORROCKS, IAN (2005). Fuzzy OWL: Uncertainty and the Semantic Web. In: International Workshop of OWL: Experiences and Directions.

STUCKENSCHMIDT, HEINER and KLEIN, MICHAEL C. A. (2003). Integrity and Change in Modular Ontologies. In: 18th international joint conference on Artificial intelligence (IJCAI'03), pp. 900–908. Morgan Kaufmann Publishers Inc.

STUCKENSCHMIDT, HEINER; PARENT, CHRISTINE; and SPACCAPIETRA, STEFANO (eds.) (2009). Modular Ontologies – Concepts, Theories and Techniques for Knowledge Modularization. Springer.

STUDER, RUDI; BENJAMINS, V. RICHARD; and FENSEL, DIETER (1998). Knowledge Engineering: Principles and Methods. Data Knowledge Engineering, 25(1–2), 161–197.

SUPEKAR, KAUSTUBH; PATEL, CHINTAN; and LEE, YUGYUNG (2004). Characterizing quality of knowledge on Semantic Web. In: 17th International FLAIRS Conference (FLAIRS'04).

SURE, YORK; STAAB, STEFFEN; and STUDER, RUDI (2003). On-To-Knowledge Methodology. In: Steffen Staab and Rudi Studer (eds.), Handbook on Ontologies, Series on Handbooks in Information Systems, pp. 117–132. Springer.

TAKEDA, HIDEAKI; IINO, KENJI; and NISHIDA, TOYOAKI (1995). Agent Organization and Communication with Multiple Ontologies. International Journal of Cooperative Information Systems, 4(4), 321–337.

TALMY, LEONARD (1983). How language structures space. In: Herbert L. Pick and Linda P. Acredolo (eds.), Spatial Orientation: Theory, Research, and Application, pp. 225–282. Plenum Press.

TALMY, LEONARD (2006). The fundamental system of spatial schemas in language. In: Beate Hampe (ed.), From Perception to Meaning: Image Schemas in Cognitive Linguistics, pp. 37–47. Mouton de Gruyter.

TANG, JIE; LI, JUANZI; LIANG, BANGYONG; HUANG, XIAOTONG; LI, YI; and WANG, KEHONG (2006). Using Bayesian decision for ontology mapping. Journal of Web Semantics, 4(4).

TARSKI, ALFRED (1959). What is Elementary Geometry? Studies in Logic and the Foundations of Mathematics, 27, 16–29.

TARTIR, SAMIR; ARPINAR, I. BUDAK; MOORE, MICHAEL; SHETH, AMIT P.; and ALEMAN-MEZA, BOANERGES (2005). OntoQA: Metric-based ontology quality analysis. In: IEEE ICDM 2005 Workshop on Knowledge Acquisition from Distributed, Autonomous, Semantically Heterogeneous Data and Knowledge Sources.

TAYLOR, MATTHEW E.; MATUSZEK, CYNTHIA; KLIMT, BRYAN; and WITBROCK, MICHAEL (2007). Autonomous Classification of Knowledge into an Ontology. In: David Wilson and Geoff Sutcliffe (eds.), 20th International FLAIRS Conference (FLAIRS'07), pp. 140–145. AAAI Press.

TELLEX, STEFANIE (2010). Natural Language and Spatial Reasoning. Ph.D. thesis, Massachusetts Institute of Technology.

TIMPF, SABINE; VOLTA, GARY S.; POLLOCK, DAVID W.; and EGENHOFER, MAX J. (1992). A Conceptual Model of Wayfinding Using Multiple Levels of Abstractions. In: Andrew U. Frank; Irene Campari; and Ubaldo Formentini (eds.), Theories and methods of spatio-temporal reasoning in geographic space, pp. 348–367. Springer.

TRASK, ROBERT L. and MAYBLIN, BILL (2000). Introducing Linguistics. Totem Books.

TVERSKY, AMOS (1977). Features of Similarity. Psychological Review, 84(4), 327–352.

UMKEHRER, ELISABETH and SCHILL, KERSTIN (1995). A general framework for comparing numerical uncertainty theories. In: 3rd International Symposium on Uncertainty Modeling and Analysis and Annual Conference of the North American Fuzzy Information Processing Society (ISUMA - NAFIPS'95), pp. 613–618.

US GSA (2007). US Courts Design Guide. Judicial Conference of the United States. US General Services Administration (GSA). April 23 2010.

USCHOLD, MIKE and GRUNINGER, MICHAEL (1996). Ontologies: Principles, methods and applications. Knowledge Engineering Review, 11(2).

VAN BENTHEM, JOHAN and BEZHANISHVILI, GURAM (2007). Modal Logics of Space. In: Marco Aiello; Ian E. Pratt-Hartmann; and Johan van Benthem (eds.), Handbook of Spatial Logics, pp. 217–298. Springer.

VAN RIJSBERGEN, CORNELIS JOOST (1979). Information Retrieval. Butterworths. `http://www.dcs.gla.ac.uk/Keith/Preface.html`, visited on July 04, 2011.

VERNON, DAVID (2008). Cognitive vision: The case for embodied perception. Image and Vision Computing, 26(1), 127–140.

VÖLKER, JOHANNA; HAASE, PETER; and HITZLER, PASCAL (2008). Learning Expressive Ontologies. In: Paul Buitelaar and Philipp Cimiano (eds.), Bridging the Gap between Text and Knowledge, pp. 45–69. IOS Press.

WALLGRÜN, JAN OLIVER; FROMMBERGER, LUTZ; WOLTER, DIEDRICH; DYLLA, FRANK; and FREKSA, CHRISTIAN (2007). A Toolbox for Qualitative Spatial Representation and Reasoning. In: Thomas Barkowsky; Markus Knauff; Gérard Ligozat; and Daniel R. Montello (eds.), Spatial Cognition V: Reasoning, Action, Interaction: International Conference Spatial Cognition 2006, pp. 79–90. Springer.

WEI, GONGJIN; BAI, WEIJING; YIN, MEIFANG; and ZHANG, SONGMAO (2010). Building the Knowledge Base to Support the Automatic Animation Generation of Chinese Traditional Architecture. In: Yaxin Bi and Mary-Anne Williams (eds.), 4th International Conference on Knowledge Science, Engineering and Management (KSEM'10), pp. 210–221. Springer.

WEIDENBACH, CHRISTOPH; DIMOVA, DILYANA; FIETZKE, ARNAUD; KUMAR, ROHIT; SUDA, MARTIN; and WISCHNEWSKI, PATRICK (2009). SPASS Version 3.5. In: 22nd International Conference on Automated Deduction (CADE'09), pp. 140–145.

WERTHEIMER, MAX (1924). Über Gestalttheorie. An address before the Kant Society, Berlin, 7th December 1924, printed in Philosophische Zeitschrift fr Forschung und Aussprache 1, pp. 39–60 (1925).

WILLIAMS, MARY-ANNE (2008). Representation = Grounded Information. In: Tu Bao Ho and Zhi-Hua Zhou (eds.), Trends in Artificial Intelligence, 10th Pacific Rim International Conference on Artificial Intelligence (PRICAI'08), vol. 5351 of *Lecture Notes in Computer Science*, pp. 473–484. Springer.

WOODS, WILLIAM A. (1975). What's in a Link: Foundations for Semantic Networks. In: Daniel G. Bobrow and Allan Collins (eds.), Representation and Understanding – Studies in Cognitive Science, Language, Thought, and Culture – Advances in the Study of Cognition, pp. 35–82. Academic Press.

WOODS, WILLIAM A. and SCHMOLZE, JAMES G. (1992). The KL-ONE family. Computers & Mathematics with Applications, 23(2-5), 133–177.

WÜNSTEL, MICHAEL (2009). 3-D Objekterkennung und Szeneninterpretation: Ein System zur multimodalen Beschreibung von Innenraumszenen. Ph.D. thesis, Universität Bremen.

YANG, YI and CALMET, JACQUES (2005). OntoBayes: An Ontology-Driven Uncertainty Model. In: International Conference on Computational Intelligence for Modelling, Control and Automation and International Conference on Intelligent Agents, Web Technologies and Internet Commerce (CIMCA-IAWTIC'06), pp. 457–463. IEEE Computer Society.

YARBUS, ALFRED L. (1967). Eye Movements and Vision. Plenum Press.

YUEN, JENNY and TORRALBA, ANTONIO (2010). A data-driven approach for event prediction. In: 11th European Conference on Computer Vision (ECCV'10), pp. 707–720. Springer.

ZADEH, LOTFI ASKER and KACPRZYK, JANUSZ (1992). Fuzzy logic for the management of uncertainty. WPC. Wiley.

ZETZSCHE, CHRISTOPH and KRIEGER, GERHARD (2001). Nonlinear mechanisms and higher-order statistics in biological vision and electronic image processing, Review and Perspective. Journal of Electronic Imaging, 10(1), 56–99.

ZIMMERMANN, H.-J. (2000). An application-oriented view of modeling uncertainty. European Journal of Operational Research, 122(2), 190–198.

ZLATEV, JORDAN (2007). Spatial Semantics. In: Dirk Geeraerts and Hubert Cuyckens (eds.), The Oxford Handbook of Cognitive Linguistics, chap. 13, pp. 318–350. OUP.